Advances in Modelling and Analysis of Functionally Graded Micro- and Nanostructures

Online at: https://doi.org/10.1088/978-0-7503-6024-1

Advances in Modelling and Analysis of Functionally Graded Micro- and Nanostructures

Edited by
Subrat Kumar Jena
Department of Applied Mechanics, Indian Institute of Technology Delhi, New Delhi, India

S Pradyumna
Department of Applied Mechanics, Indian Institute of Technology Delhi, New Delhi, India

S Chakraverty
Department of Mathematics, National Institute of Technology Rourkela, Odisha, India

IOP Publishing, Bristol, UK

ISBN 978-0-7503-6024-1 (ebook)
ISBN 978-0-7503-6022-7 (print)
ISBN 978-0-7503-6025-8 (myPrint)
ISBN 978-0-7503-6023-4 (mobi)

DOI 10.1088/978-0-7503-6024-1

Version: 20241201

IOP ebooks

British Library Cataloguing-in-Publication Data: A catalogue record for this book is available from the British Library.

Published by IOP Publishing, wholly owned by The Institute of Physics, London

IOP Publishing, No.2 The Distillery, Glassfields, Avon Street, Bristol, BS2 0GR, UK

US Office: IOP Publishing, Inc., 190 North Independence Mall West, Suite 601, Philadelphia, PA 19106, USA

May the light of Jupiter shine upon us, guiding us toward higher understanding and wisdom!

Contents

Preface

Functionally graded materials (FGMs) have attracted considerable interest in recent years because of their distinctive mechanical, thermal, and electrical characteristics. FGMs are composite materials with a gradual variation in composition and microstructure across their length. This unique characteristic allows for the customization of their properties to meet specific design criteria. Nano- and microscale functionally graded materials have been thoroughly researched because of their potential utilisation in cutting-edge technologies like micro-electro-mechanical systems (MEMS), nano-electro-mechanical systems (NEMS), biomedical devices, and others. The analysis of functionally graded nano- and microstructures, both in static and dynamical conditions, is a crucial research field with substantial implications for the design and advancement of advanced materials and structures. Understanding the dynamic behavior of functionally graded materials at the nano and microscales is critical and demanding in order to predict their mechanical response and enhance their design for specific applications.

This edited book aims to compile the recent advancements and discoveries in the field of vibration, bending, and buckling analysis of nano- and microstructures with functional grades. The book will encompass a diverse selection of subjects, such as modelling, analysis, and the practical implementation of functionally graded micro- and nanostructures across multiple domains. Graduate students, researchers, and professionals in mechanical and civil engineering, materials science, applied mathematics, and related fields are the target audience for this book. This will be a valuable resource for those interested in understanding the static and dynamic behaviours of functionally graded nano- and microstructures, as well as their applications in advanced technologies. The present book comprises a total of nine chapters, each addressing various aspects of functionally graded materials (FGMs) and their applications in nano and microstructures.

Chapter 1 investigates the thermo-mechanical interactions in a functionally graded nanobeam subjected to an instantaneous heat source. The heat transport equation is modeled using memory-dependent derivatives, and computational results are derived to analyze the effects of various parameters on field quantities such as deflection, temperature, and displacement. FG micro/nanomaterials are applied in diverse sectors, including aerospace, mining, defense, biomechanics, and manufacturing. **Chapter 2** deals with the flexural characteristics of nanobeams at the nanoscale using the mixture unified gradient theory of elasticity. An elastic short stubby nanobeam with effective material properties varying across its thickness is considered. A stationary variational framework is established, integrating all governing equations into a single functional. The boundary-value problem of equilibrium is determined and enriched with non-standard boundary conditions. The stress gradient theory, strain gradient theory, and classical elasticity theory are unified within the mixture unified gradient theory of elasticity. The flexural characteristics of the mixture unified gradient beam are addressed analytically, and a comprehensive numerical study demonstrates the flexural features of a

Timoshenko–Ehrenfest nanobeam, providing a new benchmark for numerical analysis.

Chapter 3 focuses on the size-dependent vibration analysis of functionally graded nanobeams restrained by elastic springs, using strain gradient theory. An eigenvalue problem is formulated to calculate the strain gradient theory of functionally graded nanobeams with both rigid and elastic boundary conditions. The problem is adaptable to any boundary condition without requiring re-solution. The eigenvalue problem, obtained by applying Fourier sine series and Stokes' transform, includes elastic spring stiffnesses, size parameter, and functionally graded strain gradient nanobeam properties. The study highlights how assigning small and large values to spring stiffnesses causes these springs to behave like rigid boundaries. In **Chapter 4**, the impact of micromechanical models on the behavior of functionally graded nanobeams is investigated using classical beam theory. Effective material properties of a two-phase particle composite, varying with particle volume fraction throughout the nanobeam's thickness, are determined. The study performs a free vibration analysis, considering changes in particle volume fraction. The State-Space approach and Differential Quadrature Method are used to solve partial differential equations. Numerical results show the influence of explicit micromechanical models on the fundamental frequency of the nanobeam.

Chapter 5 explores the free vibration behavior of functionally graded (FG) nanobeams made of aluminum and alumina using the Chebyshev Polynomials-based Rayleigh-Ritz method. The nanobeams have varying mechanical characteristics due to a power law distribution of volume fraction. The study uses Eringen's nonlocal elasticity theory and a single-length scale parameter within Euler-Bernoulli beam theory to model the nanobeam. A generalized eigenvalue problem is constructed to determine the non-dimensional frequency parameter. The results are compared with literature and found to be in good agreement when the dimensionless small-scale parameter is set to zero. The study also provides insights into the free vibration behavior of functionally graded nanobeams and demonstrates the effectiveness of the Chebyshev Polynomials-based Rayleigh-Ritz method. **Chapter 6** investigates the flexural vibration of functionally graded nano plates using classical plate theory and Eringen's nonlocal elasticity theory. The material properties vary spatially across thickness in a power-law form. Numerical modeling is done using the Rayleigh-Ritz method, expressing trial functions as linear combinations of algebraic polynomials. The chapter addresses the effect of different physical parameters on eigenfrequencies, validated by comparisons with available results.

Chapter 7 discusses the nonlinear static bending analysis of functionally graded micro and nanoshells, focusing on the impact of structural porosity defects. The shell geometry is analyzed in various standard geometries, including spherical, cylindrical, and conical. Moderately thick shell structures and large static deformations are considered. The strain components are obtained using the first-order shear deformation theory (FSDT). The modified couple stress theory for micro-scale analysis and nonlocal elasticity theory for nanoscale analysis are used. The shell material is functionally graded, and the material properties change along the thickness. The static governing equations and boundary conditions are derived using Hamilton's energy principle, and the static nonlinear governing equations are solved using the SAPM solution method. The influence of various factors, including shell geometry, environmental and structural factors, FGM material, boundary conditions, and small-scale analysis effects, is analyzed.

Chapter 8 emphasizes the critical fabrication procedures and design factors for the effective application of Functionally graded nano and microstructures (FGNMS) in Micro-Electro-Mechanical Systems (MEMS) and Nano-Electro-Mechanical Systems (NEMS), offering customizable mechanical, thermal, electrical, and optical functions. So this chapter provides a comprehensive review of FGNMS's transformational potential and capabilities. **Chapter 9** explores the use of FGMs in drug delivery systems, implantable devices, biosensors, and diagnostic tools. Functionally graded materials are unique structures with controlled composition and structure at nano and micro scales. They can be used in biomedical devices to customize mechanical, biological, and physicochemical properties. This chapter provides insights into their applications and impacts on healthcare systems, serving as a valuable resource for researchers, engineers, and healthcare professionals interested in using FGNMS to advance biomedical devices.

We trust that the comprehensive and challenging approaches presented in all chapters will benefit readers in their future studies and research.

Editors,
Subrat Kumar Jena, S. Pradyumna, and S. Chakraverty

Acknowledgment

The first editor, **Dr. Subrat Kumar Jena**, is immensely grateful to his family members, specifically Sh. Ullash Chandra Jena, Sh. Durga Prasad Jena, Sh. Laxmidhara Jena, Smt. Urbasi Jena, Smt. Renu Bala Jena, Smt. Arati Jena, and his sisters Jyotrimayee, Truptimayee, and Nirupama, for their unwavering love, constant motivation, unrelenting support, and blessings. Dr. Jena expresses deep indebtedness to his younger brother, Dr. Rajarama Mohan Jena, and sister-in-law, Dr. Sujata Swain, for their encouragement, love, and support. Additionally, Dr. Subrat Kumar Jena extends gratitude to his mentor Prof. Dineshkumar Harursampath of IISc Bengaluru for the moral support, faith, and encouragement. Furthermore, the first editor would like to express his deep appreciation to his Ph.D. supervisor, Prof. S. Charaverty, for his invaluable guidance and support throughout the early stages of his research. Finally, Dr. Jena would like to acknowledge the administration of the Indian Institute of Technology Delhi, and his Postdoc mentor, Prof. S. Pradyumna, for support, and encouragement.

Prof. S. Pradyumna, the second editor, is indebted to his parents Mr. Sathya Simha and Mrs. Annapoorna for their encouragement. He also expresses his gratitude to his wife Namita and children Anupam and Gayatri for their consistent support and for being source of inspiration. Pradyumna appreciates his colleagues in the Department of Applied Mechanics at the Indian Institute of Technology Delhi for their invaluable support and exceptional academic environment.

The third editor, **Prof. S. Chakraverty** expresses gratitude to his beloved parents late Sh. Birendra K Chakraborty and the late Parul Chakraborty for their blessings. He also thanks his wife, Shewli, and daughters, Shreyati and Susprihaa, for their support and inspiration during this project. The support of the NIT Rourkela administration is also gratefully acknowledged.

We, three editors, extend our heartfelt gratitude and acknowledgment to the reviewers for their invaluable feedback and appreciation during the development of the book proposal. We also express our utmost appreciation to all the chapter contributors for their timely submissions and excellent efforts. Our gratitude extends to the entire team at IOP for their unwavering support, cooperation, and assistance, which have enabled the timely publication of this book. Lastly, we express our deep indebtedness to the authors and researchers cited in the bibliography/reference sections at the end of each chapter. Their contributions have enriched this work immensely.

Editor biographies

Subrat Kumar Jena

Dr. Subrat Kumar Jena is presently working as a Postdoctoral Fellow at the Department of Applied Mechanics in the Indian Institute of Technology Delhi, New Delhi, India under the mentorship of Prof. S. Pradyumna. His academic journey includes an enriching tenure as an Honorary Postdoctoral Fellow at the Nonlinear Multifunctional Composites - Analysis & Design (NMCAD) Lab, Department of Aerospace Engineering, Indian Institute of Science (IISc), Bengaluru, India, under the mentorship of Prof. Dineshkumar Harursampath. Dr. Jena holds a PhD from the Department of Mathematics, National Institute of Technology Rourkela, Odisha, India, supervised by Prof. S. Chakraverty. His research expertise encompasses Computational Solid Mechanics, Multifunctional Materials, Applied Mathematics, Mathematical Modelling, and Uncertainty Quantification. His research contributions are recognized with 35 research papers in peer-reviewed international journals, 2 international conference papers, 9 book chapters, and 5 books to his credit. In 2023, Dr. Jena has been honoured with the "Mathematics 2022 Best PhD Thesis Award" by the Mathematics Journal, Basel, Switzerland, for his doctoral research work, marking a significant milestone in his academic journey. In addition, Dr. Jena has received an IOP Publishing Top-Cited Paper Award in 2021 and 2022 from India, published across the entire IOP Publishing journal portfolio in the past three years i.e., 2018–2020 and 2019–2021, respectively, for his paper that ranked among the top 1% of the most-cited papers in the materials subject category. In 2020-2021, one of his papers published in ZAMM – Journal of Applied Mathematics and Mechanic (Wiley) was among the most cited papers. Furthermore, three of his papers have been recognized as the best-cited papers published in the Curved and Layered Structures Journal (De Gruyter). His contributions to the field of shell buckling, particularly in the area of static or dynamic analysis of general structures, have been acknowledged, as he is included in the Shell Buckling website as "Shell Buckling People." Dr. Jena is also serving as a reviewer and guest editor for many prestigious international journals by reviewing over 100 manuscripts.

S Pradyumna

Prof. S. Pradyumna has more than 15 years of experience in academics. Presently, he is working as a Professor in the Department of Applied Mechanics at Indian Institute of Technology Delhi, India. He completed his Bachelors of Engineering in Civil Engineering from University of Mysore, Karnataka in 2000. He obtained his Master of Technology in Computer Aided Design of Structures from Visvesvaraya Technological University, Karnataka in 2003. Pradyumna received his Ph.D. from the Indian Institute of Technology Kharagpur in 2009. He joined IIT Delhi in 2010 as an Assistant Professor in the Department of Applied Mechanics. Pradyumna teaches subjects like Mechanics, Solid Mechanics, Finite Element Method, Dynamics to undergraduate and graduate students. Prior to joining IIT Delhi, he worked as an Assistant Professor in the Department of Civil Engineering at National Institute of Technology Rourkela, Odisha, India for two years (2008–2010). His areas of research interest include composite structures, functionally graded materials, plate and shell structures, computational mechanics. He has published about 44 papers in the peer reviewed national/international journals and presented 40 papers in the national/international conferences. He regularly reviews papers for reputed international journals like Composite Structures, Journal of Thermal Stresses, European Journal of Mechanics/A Solids etc.

S Chakraverty

Prof. S. Chakraverty has 30 years of experience as a researcher and teacher. Presently, he is working in the Department of Mathematics (Applied Mathematics Group), National Institute of Technology, Rourkela, Odisha, as a senior (higher administrative grade) professor. Prior to this, he was with CSIR-Central Building Research Institute, Roorkee, India. After completing graduation from St. Columba's College (Ranchi University), his career started from the University of Roorkee (now Indian Institute of Technology, Roorkee) and did M.Sc. (Mathematics) and M.Phil. (Computer Applications) from the said institute securing the first positions in the university. Dr. Chakraverty received his Ph.D. from IIT-Roorkee in 1993. Thereafter, he did his post-doctoral research at the Institute of Sound and Vibration Research (ISVR), University of Southampton, U.K., and at the Faculty of Engineering and Computer Science, Concordia University, Canada. He was also a visiting professor at Concordia and McGill universities, Canada, during 1997–1999 and visiting professor at the University of Johannesburg, Johannesburg, South Africa, during 2011–2014. He has authored/co-authored/edited 31 books, published 430 research papers (till date) in journals and conferences, and two books are ongoing. He is in the editorial boards of various international journals, book series, and conferences. Prof. Chakraverty is the chief editor of *International Journal of Fuzzy Computation and Modelling* (IJFCM), Inderscience Publisher, Switzerland (http://www.inderscience.com/ijfcm), associate editor of *Computational Methods in Structural Engineering, Frontiers in Built Environment and Curved and Layered Structures* (De Gruyter) and happens to be the editorial board member of *Springer Nature Applied Sciences, IGI Research Insights Books, Springer Book Series of Modeling and Optimization in Science and Technologies, Coupled Systems Mechanics* (Techno Press), *Journal of Composites Science* (MDPI), *Engineering Research Express* (IOP), and *Applications and Applied Mathematics: An International Journal*. He is also the reviewer of around 50 national and international journals of repute and he was the president of the section of mathematical sciences (including Statistics) of Indian Science Congress

(2015–2016) and was the vice president – Orissa Mathematical Society (2011–2013). Prof. Chakraverty is a recipient of prestigious awards, viz. Indian National Science Academy (INSA) nomination under International Collaboration/Bilateral Exchange Program (with the Czech Republic), Platinum Jubilee ISCA Lecture Award (2014), CSIR Young Scientist Award (1997), BOYSCAST Fellow. (DST), UCOST Young Scientist Award (2007, 2008), Golden Jubilee Director's (CBRI) Award (2001), INSA International Bilateral Exchange Award ([2010–2011] selected but could not undertake [2015] selected), Roorkee University Gold Medals (1987, 1988) for first positions in M.Sc. and M.Phil. (Computer Application). He is in the list of 2% world scientists (2020 and 2021) in Artificial Intelligence and Image Processing category based on an independent study done by Stanford University scientists. Prof. Chakraverty has received an IOP Publishing Top-Cited Paper Award in 2021 and 2022 from India, published across the entire IOP Publishing journal portfolio in the past three years i.e., 2018–2020 and 2019–2021, respectively, for his paper that ranked among the top 1% of the most-cited papers in the materials subject category. In 2020-2021, one of his papers published in ZAMM – Journal of Applied Mathematics and Mechanic (Wiley) was among the most cited papers. He has already guided 25 Ph.D. students and 12 are ongoing. Prof. Chakraverty has undertaken around 16 research projects as a principal investigator funded by international and national agencies totalling about Rs. 1.5 crores. He has hoisted around eight international students with different international/national fellowships to work in his group as PDF, Ph.D., visiting researchers for different periods. A good number of international and national conferences, workshops, and training programs have also been organized by him. His present research area includes differential equations (ordinary, partial, and fractional), numerical analysis and computational methods, structural dynamics (FGM, nano) and fluid dynamics, mathematical and uncertainty modelling, soft computing and machine intelligence (artificial neural network, fuzzy, interval, and affine computations).

List of contributors

Ahmed Abouelrega
Department of Mathematics, Mansoura University, Mansoura 7650001, Egypt

Naveen Kumar Akkasali
Department of Mechanical Engineering, National Institute of Technology Rourkela, Rourkela, Odisha 769008, India

S. Ali Faghidian
Department of Mechanical Engineering, Science and Research Branch, Islamic Azad University, Tehran, Iran

Rabia Benferhat
Laboratory of Geomatics and Sustainable Development, University of Tiaret, BP P 78 zaâroura, 14000 Tiaret, Algeria

S Chakraverty
Department of Mathematics, National Institute of Technology Rourkela, Rourkela, Odisha - 769008, India

Shahriar Dastjerdi
Department of Mechanical Engineering, Ferdowsi University of Mashhad, Mashhad 91775-1111, Iran

Kada Draiche
Civil Engineering Department, University of Tiaret, BP P 78 zaâroura, 14000 Tiaret, Algeria

Akash Kumar Gartia
Department of Mathematics, National Institute of Technology Rourkela, Rourkela, Odisha - 769008, India

Mehran Kadkhodayan
Department of Mechanical Engineering, Ferdowsi University of Mashhad, Mashhad 91775-1111, Iran

Erukala Kalyan Kumar
Department of Mechanical Engineering, National Institute of Technology Rourkela, Rourkela, Odisha 769008, India

Vikash Kumar
Department of Mechanical Engineering, National Institute of Technology Rourkela, Rourkela, Odisha 769008, India

Ashish Kumar Meher
Department of Mechanical Engineering, National Institute of Technology Rourkela, Rourkela, Odisha 769008, India

Santwana Mukhopadhyay
Department of Mathematical Sciences, Indian Institute of Technology (BHU), Varanasi, Uttar Pradesh 221005, India

Subrata Kumar Panda
Department of Mechanical Engineering, National Institute of Technology Rourkela, Rourkela, Odisha 769008, India

K K Pradhan
Department of Mathematics, Government Women's College, Sundargarh, Odisha - 770001, India

Hassaine Daouadji Tahar
Laboratory of Geomatics and Sustainable Development, University of Tiaret, BP P 78 zaâroura, 14000 Tiaret, Algeria

Rakhi Tiwari
Department of Mathematics, Nitishwar College, Babasaheb Bhimrao Ambedkar Bihar University, Muzaffarpur, Bihar 842002, India

Youcef TLIDJI
Materials and Structures Laboratory, Civil Engineering Department, University of Tiaret, BP P 78 zaâroura, 14000 Tiaret, Algeria

Büşra Uzun
Department of Civil Engineering, Bursa Uludag University, Gorukle Campus, 16059 Nilüfer/BURSA, Turkey

Mustafa Özgür Yaylı
Department of Civil Engineering, Bursa Uludag University, Gorukle Campus, 16059 Nilüfer/BURSA, Turkey

Chapter 1

Functionally graded micro and nanostructures: introduction, modeling, and applications

Rakhi Tiwari, Santwana Mukhopadhyay and Ahmed Abouelregal

Functionally graded (FG) materials are recently developed composite materials that reflect very interesting characteristics that their properties alter gradually through the geometry of the structures. FG materials therefore preserve the strength of the conventional composite materials and eliminate their weaknesses. Such materials are constructed by changing the chemical compositions, or design attributes from one end to the other end as necessary. The present study speculates the coupled thermomechanical interactions inside a functionally graded nanobeam affected from a sudden heat input. Modeling of the problem is established by considering the heat transport equation in terms of the memory dependent time derivatives. Based on the closed form outcomes, computational results are derived that execute the influences of the various parameters on the field quantities such as deflection, temperature, and displacement. Further, applications of the FG micro/nanomaterials in different areas such as aerospace, mining, defence, biomechanics, and manufacturing sectors are expressed.

1.1 Introduction

Functionally graded (FG) materials are attracting huge interest in industrial applications such as automotive, aerospace, biomedical implants, energy absorbing structures, optoelectronic equipment, and geological models, etc. Basically, functionally graded materials are composed of materials having smaller variations in their compositions all over the volume [1, 2] (figure 1.1).

Scientists have observed that some FG materials are commonly found in nature, e.g. tissue alterations in seashells (Cypraecassis rufa, pearly oyster, Peristernia incarnate) and changing trabecular materials. Some plants (Norway spruce and

doi:10.1088/978-0-7503-6024-1ch1

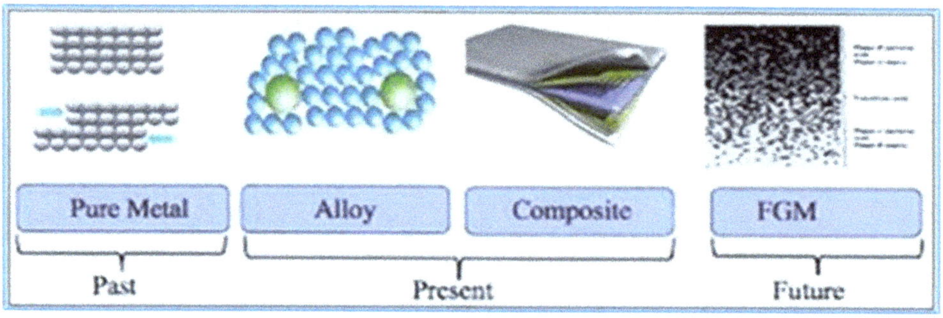

Figure 1.1. Structure of FG materials.

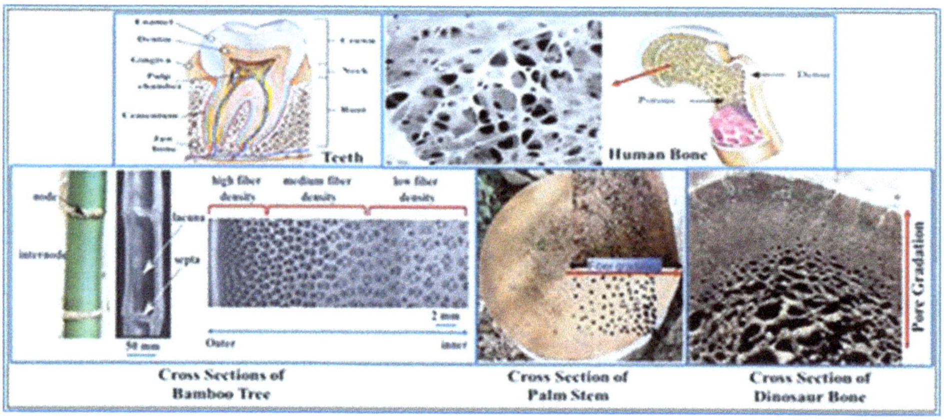

Figure 1.2. Examples of natural FG materials.

bamboo) are referred to as FG substances [3–5]. Some natural FG materials are presented in figure 1.2.

In order to investigate the structures of functionally graded materials, credit goes to Niino *et al* [6] who have invented fabricating a thermally graded metal to ceramic phase. After that, based on this invention, FG materials have been extensively investigated. In contrast to the isotropic materials, FG materials are precisely designed to exhibit numerous functional properties. This fact can be understood in such a way that advanced composite structures (FG) are fabricated from two or more integrands that differ in physical as well as chemical properties in contrast to a single substance. A mixture of two or more materials, depending on the requirements, creates several characteristics inside the composite material such as anisotropy, heterogeneity, symmetry, and hierarchy. Composite materials therefore exhibit several good properties including high strength, greater resistance to fatigue, greater reliability, wear, and corrosion, etc.

Apart from this, these structures introduce gradient interface in contrast to the sharp interface [7]. That is why the mechanical properties of the FG material such as shear modulus, Young's modulus, Poisson's ratio, density of the material, and thermal expansion coefficient, etc, alter in the smooth pattern in a continuous manner in the selected directions.

In the modern era, micro/nanostructures are of great interest among researchers due to their peculiar properties such as that they are easy to carry etc. Microstructures can have different shapes such as beams, circular disks, square plates, annular rings, etc, and can be categorized according to their modes of operation, like torsional, flexural, and bulk mode devices.

The present chapter discusses the functioning of the FG nanobeam when it is exposed to a sudden heat input at its boundary.

Functionally graded nanobeams are highly utilized for fabricating nanoelectromechanical systems which shorten their crack and thermal resistance, originated inside them. Ye *et al* [8] solved a coupled thermomechanical axisymmetric issue for a FG cylindrical cell. El Naggar *et al* [9] evaluated temperature based stress for an orthotropic rotating cylinder. Shao *et al* [10] studied coupled thermomechanical interactions generated inside the functionally graded circular shaped hollow cylinder. Analysis on thermoelastic vibration response of FG nanobeam has been reported by Abouelregal and Tiwari [11]. Performance of a FG nanobeam induced by a sinusoidal pulse heating due to fractional order heat conduction process can be understood by the work of Abouelregal and Mohamed [12].

1.2 Modeling of the problem concerned with a FG nanobeam subjected to an instant heat input

Consider a thermoelastic functionally graded nanobeam kept initially at temperature T_0. The axial direction of the beam is assumed to be x axis while the other two axes (y, z axes) are set as beam's width and thickness, respectively. The geometry of the problem is displayed in figure 1.3.

Following the characteristics of the FG materials, the properties of the beam, e.g. thermal conductivity, coupling parameters, mass density, and modulus of elasticity varies slowly but continuously along the z axes of the nanobeam.

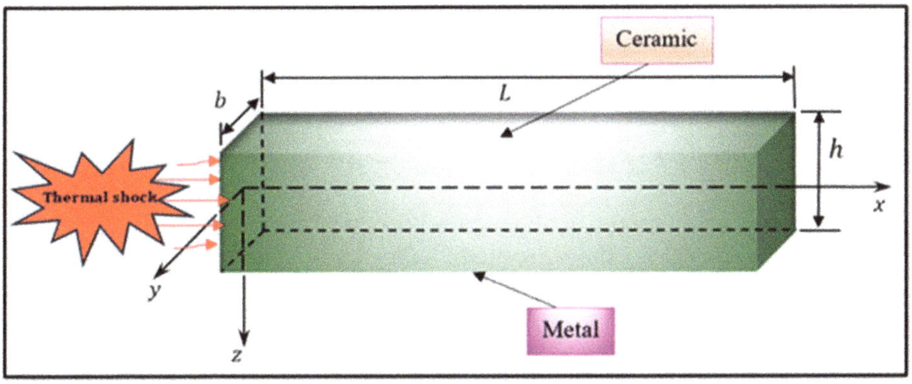

Figure 1.3. Representation of FG nanobeam.

Material gradation property $P'(z)$ along the direction of the thickness is demonstrated as [11]:

$$P'(z) = P_m e^{n_p(2z-h)/h}, \quad n_p = \ln \sqrt{P_m/P_c}, \tag{1.1}$$

where P_m and P_c denote the corresponding property of the metal and ceramic components of the FG material. It should be noted that the nanobeam comprises metallic properties at its lower surface $z = h/2$ and ceramic properties at the upper surface $z = -h/2$.

The beam exhibits bending vibrations having small amplitudes about the axial direction, hence its functioning can be modeled by adopting the postulates of the Euler–Bernoulli beam theory. We will use standard notations for the symbols throughout the document.

Displacement of the beam is therefore modeled as:

$$u = -z\frac{\partial w}{\partial x}, \quad v = 0, \quad w(x, y, z, t) = w(x, t) \tag{1.2}$$

where w represents the lateral deflection.

In addition to equations (1.1) and (1.2), the following form of heat transport equation is considered (in the absence of an external source) [13]:

$$\begin{gathered}(1 + \tau_T D_\omega)K_m e^{n_k(2z-h)/h}\left[\frac{\partial^2\theta}{\partial x^2} + \frac{\partial^2\theta}{\partial z^2} + \frac{2n_K}{h}\frac{\partial\theta}{\partial z}\right] = \\ (1 + \tau_q D_\omega)\left[\rho_m C_{Em} e^{n_{\rho C_E}(2z-h)/h}\frac{\partial\theta}{\partial t} - z\gamma_m e^{n_\gamma(2z-h)/h}T_0\frac{\partial}{\partial t}\left(\frac{\partial^2 w}{\partial x^2}\right)\right].\end{gathered} \tag{1.3}$$

An advanced definition of derivatives (memory dependent definition of derivatives) has been used in the above equation. Some recent work on generalized thermoelasticity based on the concept memory dependent derivative can be found in references [14–17].

The parameters n_k, n_γ and $n_{\rho C_E}$ in equation (1.3) are determined in terms of the properties of the materials, and

$$\gamma_m = \frac{E_m\alpha_m}{1 - 2\nu_m}, \quad \rho_m C_{Em} = \frac{K_m}{\chi_m}.$$

Further note that in equation (1.3), the definition of the first order derivative comprising memory effect is evaluated as:

$$D_\omega f(t) = \frac{1}{\omega}\int_{t-\omega}^{t} K(t-p)f'(p)dp.$$

ω reflects the time-delay quantity. The kernel function $K(t-p)$ is chosen according to the requirements of the problem. However, its range is defined as $0 < K(t-p) \leqslant 1$ for $p \in [t-\omega, t]$. It is noted that the memory based derivative exhibits a smaller quantity in contrast to the normal derivative.

If $K(t-p) = 1$ and $\omega \to 0$, we obtain

$$D_\omega f(t) = \frac{1}{\omega}\int_{t-\omega}^{t} f'(p)dp = \frac{f(t) - f(t-\omega)}{\omega} \to f'(t).$$

The memory based derivative of nth order for the function $f(t)$ is defined as

$$D^n_\omega f(t) = \frac{1}{\omega}\int_{t-\omega}^{t} K(t-p)f^n(p)dp.$$

Hence, the heat transport equation in terms of memory based derivatives is determined as:

$$K(1+\tau_T D_\omega)\nabla^2\theta = (1+\tau_q D_\omega)\left(\rho C_v\dot{\theta} + \gamma T_0\frac{\partial}{\partial t}(\mathrm{div}(u))\right).$$

For the present study, the following kernel functions are chosen as [11, 13–15]

$$K(t-p) = 1 - \frac{2b}{\omega}(t-p) + \frac{a^2}{\omega^2}(t-p)^2 = \begin{cases} K_1 = \left(\frac{p-t}{\omega}+1\right)^2; \ a=b=1, \\ K_2 = \left(\frac{p-t}{\omega}+1\right); \ a=0,\ b=\frac{1}{2}, \\ K_3 = 1; \ a=b=0, \end{cases} \quad (1.4)$$

where, K_3 demonstrates the case of the traditional definition of the derivative in place of a memory based derivative.

The beam is assumed to be insulated so that no heat flow is found across its upper and lower surfaces. Hence, $\frac{\partial\theta}{\partial z}=0$ vanishes for $z=\pm h/2$.

Being extremely thin, it is considered that the temperature increment alters in a sinusoidal form along the direction of the thickness of the beam.

Then,

$$\theta(x, z, t) = \Theta(x, t)\sin\left(\frac{\pi z}{h}\right). \quad (1.5)$$

Substituting equation (1.5) into equation (1.3) and integrating the final equation w.r.t. z and taking the limits $-h/2$ to $h/2$, we gather the following equation:

$$(1+\tau_T D_\omega)\frac{\partial^2\Theta}{\partial x^2} = (1+\tau_q D_\omega)\left[\overline{\mu}_{\rho C_E}\eta\frac{\partial\Theta}{\partial t} - \frac{\overline{\mu}_\gamma\gamma_m hT_0}{K_m}\frac{\partial}{\partial t}\left(\frac{\partial^2 w}{\partial x^2}\right)\right], \quad (1.6)$$

where, $\eta = \rho_m C_{Em}/K_m$, $\overline{\mu}_{\rho C_E} = \mu_{\rho C_E}/\mu_K$ and $\overline{\mu}_\gamma = \mu_\gamma/\mu_K$,

with

$$\mu_{\rho C_E} = \frac{2n_{\rho C_E}(1+e^{-2n_{\rho C_E}})}{\pi^2+4(n_{\rho C_E})^2},\quad \mu_K = \frac{2n_K(1+e^{-2n_K})}{\pi^2+4(n_K)^2},\quad \mu_\gamma = \frac{n_\gamma(1+e^{-2n_\gamma})+e^{-2n_\gamma}-1}{4(n_\gamma)^2}.$$

In the present study, the nonlocal elasticity theory is considered to study the small sized beam structure and the equation of motion in the present context is given by

$$\sigma_x - \xi\frac{\partial^2\sigma_x}{\partial x^2} = -E_m\left[ze^{\frac{n_E(2z-h)}{h}}\frac{\partial^2 w}{\partial x^2} + \alpha_m\theta e^{\frac{n_{E\alpha}(2z-h)}{h}}\right], \quad (1.7)$$

where $n_{E\alpha} = \ln\sqrt{E_m\alpha_m/E_c\alpha_c}$ in which E_c, α_c are Young's modulus and the thermal expansion coefficient of the ceramic material, respectively. ξ denotes here the nonlocal parameter.

Using equation (1.7), the cross sectional flexure moment is defined as

$$M(x,\,t) - \xi\frac{\partial^2 M}{\partial x^2} = -bh^2E_m\left[h\mu_E\frac{\partial^2 w}{\partial x^2} + \alpha_m\mu_{E\alpha}\,\Theta\right], \tag{1.8}$$

where

$$\mu_E = \frac{\left(n_E^2+2\right)(1-e^{-2n_E}) - 2n_E(1-e^{-2n_E})}{8n_E^3},$$

$$\mu_K = \frac{2n_{E\alpha}\left(\pi^2+4n_{E\alpha}^2\right)(1-e^{-2n_{E\alpha}}) + \left(\pi^2-4n_{E\alpha}^2\right)(1+e^{-2n_{E\alpha}})}{\left(\pi^2+4n_{E\alpha}^2\right)^2}.$$

The equation of transverse motion is stated as [14, 15]:

$$\frac{\partial^2 M}{\partial x^2} = \frac{(1-e^{-2n_\rho})\rho_m}{2n_\rho}A\frac{\partial^2 w}{\partial t^2}. \tag{1.9}$$

Implementing equations (1.8) into (1.9), the equation of motion is obtained as:

$$\frac{\partial^4 w}{\partial x^4} + \frac{\rho_m(1-e^{-2n_\rho})}{2E_mh^2n_\rho\mu_E}\left(\frac{\partial^2 w}{\partial t^2} - \xi\frac{\partial^4 w}{\partial t^2\partial x^2}\right) + \frac{\alpha_{m\mu_{E\alpha}}}{\mu_E}\frac{\partial^2\Theta}{\partial x^2} = 0. \tag{1.10}$$

The flexure moment is determined by [11]

$$M(x,\,t) = \xi A\frac{(1-e^{-2n_\rho})\rho_m}{2n_\rho}\frac{\partial^2 w}{\partial t^2} - bh^2E_m\left[h\mu_E\frac{\partial^2 w}{\partial x^2} + \alpha_m\mu_{E\alpha}\Theta\right]. \tag{1.11}$$

For the purpose of simplifying the governing equations stated above, the following nondimensional entities are taken into account (see [11] for details):

$$\begin{aligned}&\left\{x',\,z',\,u',\,w',\,L',\,h'\right\} = c_0\eta_0\left\{x,\,z,\,u,\,w,\,L,\,h\right\},\ \Theta' = \frac{\Theta}{T_0},\\ &\left\{t',\,\tau'_0,\,\xi',\,\omega',\,\tau'_T,\,\tau'_q\right\} = c_0^2\eta_0\left\{t,\,\tau_0,\,\xi,\,\omega,\,\tau_T,\,\tau_q\right\}.\end{aligned} \tag{1.12}$$

Therefore, we achieve the following dimensionless equations:

$$\begin{aligned}&\frac{\partial^4 w}{\partial x^4} + G_1\left(\frac{\partial^2 w}{\partial t^2} - \xi\frac{\partial^4 w}{\partial t^2\partial x^2}\right) = -G_2\frac{\partial^2\Theta}{\partial x^2},\\ &(1+\tau_TD_\omega)\frac{\partial^2\Theta}{\partial x^2} = (1+\tau_qD_\omega)\left[G_3\frac{\partial\Theta}{\partial t} - G_4\frac{\partial}{\partial t}\left(\frac{\partial^2 w}{\partial x^2}\right)\right],\end{aligned} \tag{1.13}$$

$$M(x, t) = G_1\left(\xi\frac{\partial^2 w}{\partial t^2} - \frac{\partial^2 w}{\partial x^2}\right) - G_2\Theta. \tag{1.14}$$

$$G_1 = \frac{(1 - e^{-2n_\rho})}{2h^2 n_\rho \mu_E},\ G_2 = \frac{T_0\alpha_{m\overline{\mu}_{E\alpha}}}{h},\ G_3 = \overline{\mu}_{\rho C_E},\ G_4 = \frac{\overline{\mu}_\gamma \gamma_m h}{\eta_0 K_m}.$$

1.3 Analytical solutions

The initial conditions are taken as

$$\Theta(x, 0) = \frac{\partial\Theta(x, 0)}{\partial t} = 0 = w(x, 0) = \frac{\partial w(x, 0)}{\partial t}. \tag{1.15}$$

Adopting the Laplace transform mechanism to equations (1.13) and (1.14), we find

$$\left(\frac{d^4}{dx^4} - \xi G_1 s^2\frac{d^2}{dx^2} + G_1 s^2\right)\overline{w} = -G_2\frac{d^2\overline{\Theta}}{dx^2}, \tag{1.16}$$

$$\frac{d^2\Theta}{dx^2} = q\left[G_3\overline{\Theta} - G_4\frac{d^2\overline{w}}{dx^2}\right], \tag{1.17}$$

$$\overline{M}(x, t) = G_1\left(\xi s^2\overline{w} - \frac{d^2\overline{w}}{dx^2}\right) - G_2\overline{\Theta}, \tag{1.18}$$

where $q = \frac{(1 + \frac{\tau_T}{\omega}G(\omega))}{(1 + \frac{\tau_q}{\omega}G(\omega))}$, $G(\omega) = 1 - \frac{2b}{\omega} + \frac{2a^2}{\omega^2} - e^{-s\omega}[(1 - 2b + a^2) + \frac{2(a^2 - b)}{\omega s} + \frac{2a^2}{\omega^2 s^2}]$.

Eliminating $\overline{\Theta}$ or $\overline{w}$ from equations (1.16) and (1.17), the following equation is obtained:

$$(D^6 - E'^{D^4} + F'^{D^2} - G')\{\overline{\Theta}, \overline{w}\}(x) = 0, \tag{1.19}$$

where

$$E' = \xi G_1 s^2 + qG_3 + qG_2G_4,\ F' = G_1 s^2 + q\xi\ G_1G_3 s^2,\ G' = qG_1G_3 s^2,\quad D = \frac{d}{dx}.$$

Equation (1.19) is improved as:

$$\left(D^2 - m_1^2\right)\left(D^2 - m_2^2\right)\left(D^2 - m_3^2\right)\{\overline{\Theta}, \overline{w}\}(x) = 0, \tag{1.20}$$

where m_n^2, $n = 1, 2, 3, 4$ present the roots of equation

$$m^6 - Em^4 + Fm^2 - G = 0. \tag{1.21}$$

We therefore achieve the analytical solutions of deflection and temperature fields by solving equation (1.20) as:

$$\{\overline{w}, \overline{\Theta}\}(x) = \sum_{n=1}^{3}\{1, E_n'\}(E_n e^{-m_n x} + E_{n+3}e^{m_n x}). \tag{1.22}$$

The compatibility between equations (1.16) and (1.17), gives

$$E_n' = -\frac{m_n^4 + B_1 s^2}{B_2 m_n^2} = \beta_n E_n.$$

The axial displacement is determined as:

$$\overline{u}(x) = -z\frac{d\overline{w}}{dx} = z\sum_{n=1}^{3} m_n(E_n e^{-m_n x} - E_{n+3}e^{m_n x}). \tag{1.23}$$

Using the closed-form solutions of $\overline{w}$ and $\overline{\Theta}$ from (1.22) into (1.18), bending moment $\overline{M}$ is achieved as:

$$\overline{M}(x) = \sum_{n=1}^{3}\left(\xi s^2 - m_n^2 - B_2\beta_n\right)(E_n e^{-m_n x} + E_{n+3}e^{m_n x}). \tag{1.24}$$

Furthermore, the strain component is found as:

$$\overline{e}(x) = \frac{d\overline{u}}{dx} = -z\sum_{n=1}^{3} m_n^2(E_n e^{-m_n x} + E_{n+1}e^{m_n x}). \tag{1.25}$$

1.4 Application

Mechanical boundary conditions:

Both ends of the nanobeam are taken to be simply-supported. Therefore we consider

$$w(0, t) = w(L, t) = 0 = \frac{\partial^2 \mathrm{w}(0, \mathrm{t})}{\partial x^2} = \frac{\partial^2 \mathrm{w}(L, \mathrm{t})}{\partial x^2}. \tag{1.26}$$

Thermal boundary conditions:

The first boundary of the beam is exposed by an instantaneous heat source. Hence, the first thermal condition is defined by

$$\Theta(0, t) = q_0 H(t) \text{ on } x = 0, \tag{1.27}$$

where q_0 denotes constant intensity and $H(t)$ represents the unit step function.

The beam's second end $x = L$ is supposed to be thermally insulated. Therefore,

$$\frac{\partial\Theta}{\partial x} = 0 \text{ on } x = L. \tag{1.28}$$

Using the Laplace transform technique into equations (1.26)–(1.28), we achieve

$$\begin{aligned}
\overline{\mathrm{w}}(0, \mathrm{s}) = \overline{\mathrm{w}}(L, \mathrm{s}) = 0, \\
\frac{\partial^2 \overline{\mathrm{w}}(0, \mathrm{s})}{\partial t^2} = \frac{\partial^2 \overline{\mathrm{w}}(L, \mathrm{s})}{\partial t^2} = 0, \\
\overline{\Theta}(0, s) = \frac{q_0}{s} = P(s), \\
\frac{\partial \overline{\Theta}(L, \mathrm{s})}{\partial x} = 0.
\end{aligned} \tag{1.29}$$

Using equation (1.22) into the boundary conditions (equations (1.26)–(1.28)), the following linear equations are deduced:

$$\begin{aligned}
\sum_{n=1}^{3} (E_n + E_{n+1}) = 0, \\
\sum_{n=1}^{3} (E_n e^{-m_n L} + E_{n+1} e^{m_n L}) = 0.
\end{aligned} \tag{1.30}$$

$$\begin{aligned}
\sum_{n=1}^{3} m_n^2 (E_n + E_{n+1}) = 0, \\
\sum_{n=1}^{3} m_n^2 (E_n e^{-m_n L} + E_{n+1} e^{m_n L}) = 0.
\end{aligned} \tag{1.31}$$

$$\begin{aligned}
\sum_{n=1}^{3} m_n (\beta_n E_n - \beta_{n+1} E_{n+1}) = -P(s), \\
\sum_{n=1}^{3} m_n (\beta_n E_n e^{-m_n L} - \beta_{n+1} E_{n+1} e^{m_n L}) = 0.
\end{aligned} \tag{1.32}$$

The above system speculates the solutions for the unknown parameters E_n, $(n = 1, 2, \dots, 6)$.

1.5 Analysis of the numerical results

Here, the upper surface is assumed to be fabricated from alumina while the lower surface is considered to be constructed from aluminum. The properties of these constituents are given in table 1.1.

Additionally, the ratio of the length-to-thickness is taken as $L/h = 10$ and $z = \frac{h}{3}$, $L = 1$.

Considering the above data, computational work is performed from the theoretical solution as obtained in the previous sections and numerical results are illustrated in various graphs at time $t = 0.05$ to analyze the effects of various parameters on the field variables like, deflection, temperature, and displacement. By using the above

Table 1.1. Values of the parameters for FG nanobeam ($T_0 = 293$ K) [12].

Material Material properties	Metal (aluminum)	Ceramic (alumina)
Young's modulus E (GPa)	70	116
Thermal expansion α_c (K^{-1})	23.1×10^{-5}	8.7×10^{-6}
Thermal conductivity K (W m^{-1} K^{-1})	237	1.78
Thermal diffusivity χ (m^2 s^{-1})	84.18×10^{-6}	1.06×10^{-6}
Density ρ (kg m^{-3})	2700	3000
Poisson's ratio ν	0.35	0.33

data, numerical computation is carried out from the theoretical results obtained in the previous section and the effects of various parameters on the deflection, temperature, and displacement are illustrated in various graphs as given below.

1.5.1 Impact of nonlocal parameter

The present problem tackles the thermomechanical interaction inside the nanobeam affected by a sudden heat input; the conventional thermal conductivity model fails to provide satisfactory outcomes. Therefore, advanced nonlocalized thermoelastic theory with two relaxation times including memory based derivatives is employed here. In this subsection, the influences of the nonlocal quantity $\overline{\xi}$ are recorded on the alterations of field variables deflection w, temperature θ, and displacement u. We have considered four distinct values of nonlocal quantity $\overline{\xi} = 0.0,\ 0.1,\ 0.3,\ 0.5$ where $\overline{\xi} = 0$ reveals the absence of nonlocal character, i.e., the case of classical elasticity theory. Graphs are plotted for the nonlinear kernel function $K_1 = \left(\frac{p-t}{\omega} + 1\right)^2$. Further, time delay $\omega = 0.001$ and time relaxation parameters $\tau_q = 0.02$, $\tau_T = 0.01$ are considered.

Figure 1.4 displays the variations of the distributions of deflection quantity w. It can be noticed that the trend of variation is found to be unchanged from the nonlocal quantity $\overline{\xi}$ but the values of deflection are significantly altered as the nonlocal quantity changes its value. For $\overline{\xi} = 0$, i.e. classical elasticity theory predicts the greatest values of the deflection field especially at the peak point; however, as the nonlocal entity $\overline{\xi}$ enhances from zero, the deflection field begins to decrease. Moreover, the nonlocal quantity $\overline{\xi} = 0.5$, speculates the lowest values of the deflection quantity.

It is concluded that the nonlocalized system forecasts finite behavior of thermoelastic waves and less energy dissipation of waves. Low energy dissipation creates low thermal stress inside the micro/nanostructures and enhances their lifespan.

Figure 1.5 demonstrates the variation of temperature distribution against the beam length. Following a similar pattern to the deflection quantity, the temperature field is observed to be hugely influenced by the nonlocal parameter. Curves start from a constant value at the boundary and decreases with the increase of distance

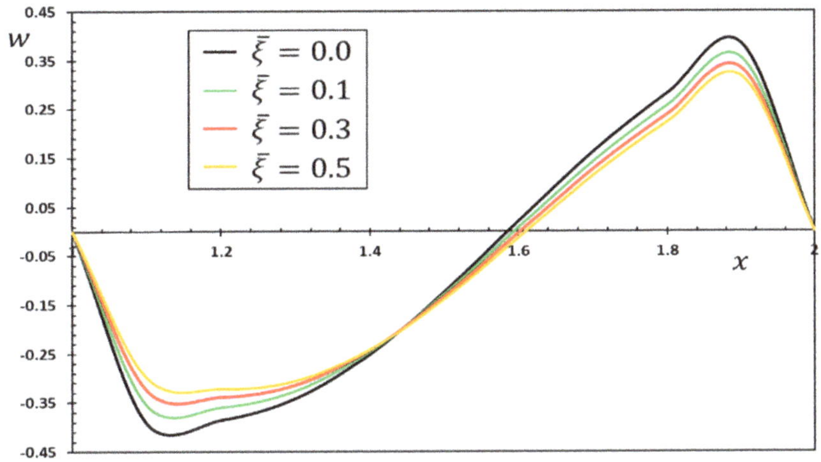

Figure 1.4. Deflection, w versus distance, x for distinct values of the nonlocal parameter.

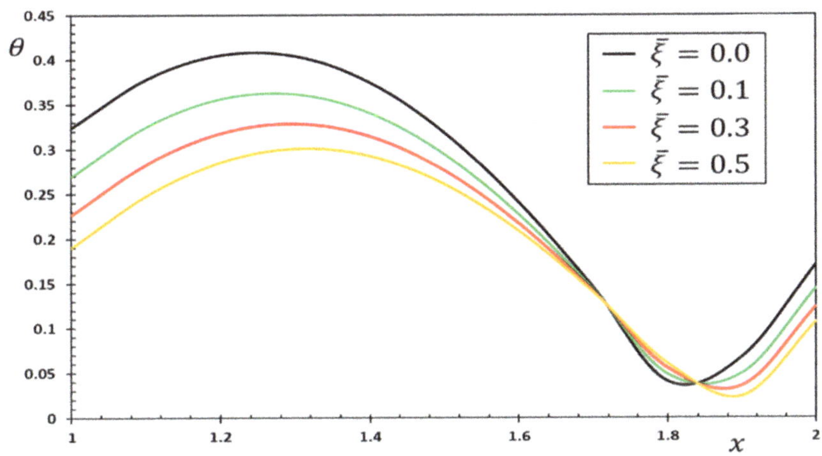

Figure 1.5. Temperature, θ versus distance, x for distinct values of the nonlocal parameter.

and disappear after passing some distance. This indicates a finite speed of thermal waves inside the FG materials nanobeam. Clearly, due to the involvement of the nonlocal quantity $\bar{\xi}$, the numerical values of temperature gets reduced.

Figure 1.6 demonstrates the impact of nonlocal parameter $\bar{\xi}$ on the displacement field. In contrast to the deflection and temperature fields, the displacement field is observed to be highly influenced by the presence of nonlocal parameter $\bar{\xi}$ near the boundary.

It is believed that these outcomes may be beneficial for scientists and engineers designing the promising structures of micro/nanomachines (figures 1.5 and 1.6).

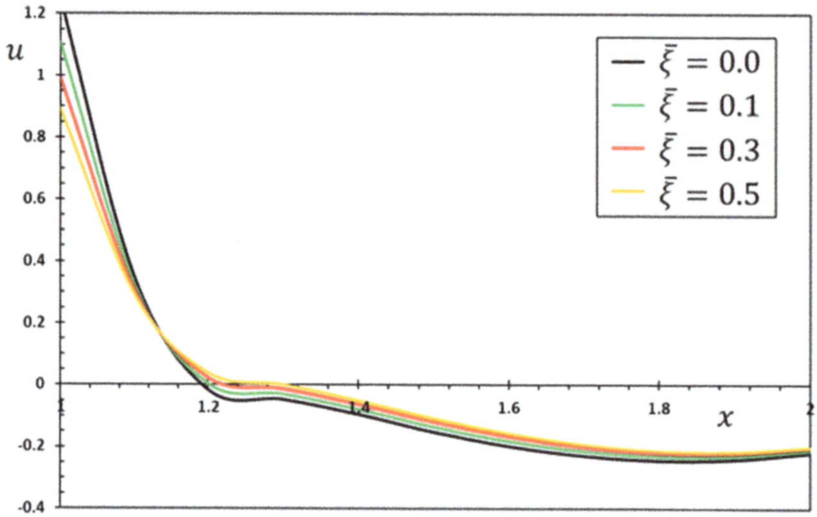

Figure 1.6. Displacement, u versus distance, x for distinct values of the nonlocal parameter.

1.5.2 Impact of kernel functions

This subsection is aimed at forecasting the influences of the considered three kernel functions $K_1 = \left(\frac{p-t}{\omega} + 1\right)^2$, $K_2 = \left(\frac{p-t}{\omega} + 1\right)$, $K_3 = 1$ (nonlinear, linear, and constant in behavior, respectively), on the alterations of the field quantities—lateral deflection w, temperature θ, displacement u versus distance x. The values of the nonlocal entity $\bar{\xi} = 2$, $\omega = 0.001$ are taken into account.

Figure 1.7 displays the nature of variation of deflection for various kernel functions. The effects of kernel functions are observed to be insignificant on the trend of variations of the deflection field. However, the deflection field is prominently affected with respect to the kernel function. Each plot exhibits a similar value at the boundary but due to the presence of kernel functions, the value of the deflection field alters between the boundaries. The peak points of the curves suggest the highest effects of the kernel function. The magnitudes of the deflection distributions are achieved to be lowest for nonlinear kernel function K_1 while the constant kernel function K_3 deduces highest values. Hence, it can be seen that the deflection field shows the greatest values for the constant kernel function where the conventional derivatives are used in place of memory based derivatives. Therefore, it is summarized that coupled thermomechanical waves move with the greater speed in the case of traditional derivatives and provide greater loss of energy inside the micro/nanostructure.

Figure 1.8 derives the impact of kernel functions on the variations of the temperature field. It is noticed that curves predict the least differences for $0 \leqslant x \leqslant 0.1$. Later on, the differences among the temperature curves enhance for the high values of the distance and finally all curves vanish after crossing a finite distance. Curves exhibit jumps and the effects of the kernel functions are

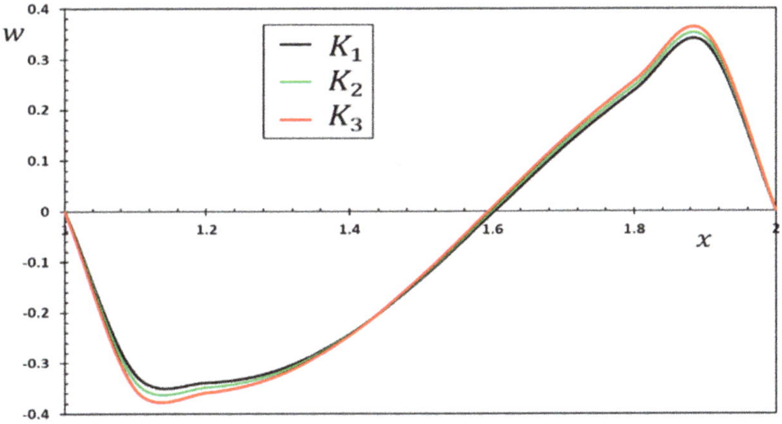

Figure 1.7. Deflection, w versus distance, x for different kernel functions.

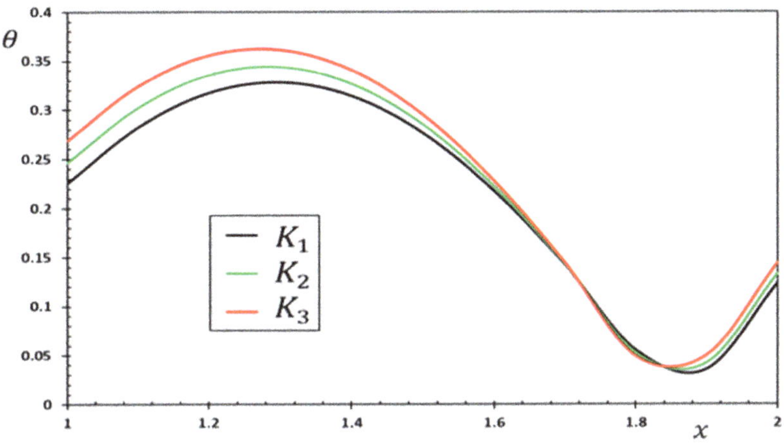

Figure 1.8. Temperature, θ versus distance, x for different kernel functions.

achieved to be highest at the jump. Similar to the deflection curves, temperature curves reveal the lowest values for the nonlinear kernel function and greatest values for the constant kernel function. Hence, it is noticed that memory dependent derivatives limit the speed of wave propagation inside the beam and provides safety from the origination of thermal stress inside the structure.

Figure 1.9 presents the displacement field alterations versus distance x. The influences of the kernel functions are observed to be least on the displacement field when compared to the impacts of the kernel functions on temperature and deflection fields. However, displacement plots reveal lower values for the linear and nonlinear kernel functions K_1, K_2.

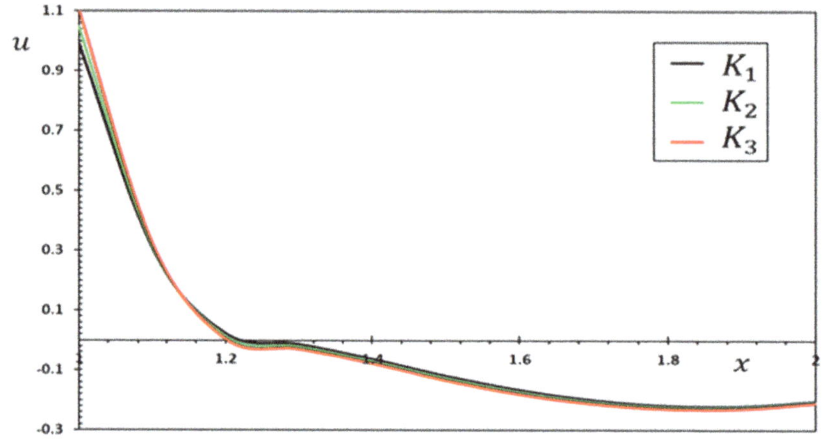

Figure 1.9. Displacement, u versus distance, x for different kernel functions.

1.6 Concluding remarks

This study presents mathematical modeling on thermoelastic interactions in functionally graded micro/nanobeams due to sudden input of heat effects by employing coupled thermoelasticity theory based on memory-dependent heat transfer equation and highlights the following conclusions:

- Physical fields, like deflection, displacement, and temperature disappear after crossing some distance from the first boundary, implying a finite speed of thermal wave propagation in the present context.
- The presence of memory dependent derivatives forecasts the lower thermal stress and consequently low energy dissipation and, in this way, this theory supports fabrication of the micro/nano FGM machines with high stability and enhances the lifespan of the machines.
- Nonlinear kernel function of memory-dependent time derivatives in heat transfer equation indicates the best performance of the nanobeam.
- Nonlocal quantity reduces the domain of influences of the physical fields and consequently reveals promising results.

The authors believe that the current investigation will be advantageous for the scientists and engineers for fabricating the optimum design and structures of FGM based machines and equipments.

Acknowledgment

Authors are thankful to the reviewers for the constructive comments. One of the authors (SM) also thankfully acknowledges the financial support from SERB, India under the MATRICS project scheme (project number: MTR/2022/000333).

References

[1] Yang N, Hu S, Ma D, Lu T and Li B 2015 Nanoscale graphene disk: a natural functionally graded material–how is Fourier's law violated along radius direction of 2D disk *Sci. Rep.* **5** 14878

[2] Loh G H, Pei E, Harrison D and Monzón M D 2018 An overview of functionally graded additive manufacturing *Addit. Manuf.* **23** 23–34

[3] Wegst U G K, Bai H, Saiz E, Tomsia A P and Ritchie R O 2014 Bio inspired structural materials *Nat. Mater.* **14** 23

[4] Eder M, Jungnikl K and Burgert I 2009 A close-up view of wood structure and properties across a growth ring of Norway spruce *Trees* **23** 79

[5] Habibi M K, Samaei A T, Gheshlaghi B, Lu J and Lu Y 2015 Asymmetric flexural behavior from bamboo's functionally graded hierarchical structure: underlying mechanisms *Acta Biomater.* **16** 178

[6] Niino M, Kisara K and Mori M 2005 Feasibility study of FGM technology in space solar power systems (SPSS) *Mater. Sci. Forum* **492** 163–8

[7] Bhavar V, Kattire P, Thakare S, patil S and Singh R K P 2017 A review on functionally gradient materials (FGMs) and their applications *IOP Conf. Ser.: Mater. Sci. Eng.* **229** 012021

[8] Ye G R, Chen W Q and Cai J B 2001 A uniformly heated functionally graded cylindrical shell with transverse isotropy *Mech. Res. Commun.* **28** 535–42

[9] El-Naggar A M, Abd-Alla A M, Fahmi M A and Ahmed S M 2002 Thermal stresses in a rotating non-hokogeneous orthotropic hollow cylinder *Heat Mass Transf.* **39** 41–6

[10] Shao Z S, Wang T J and Ang K K 2007 Transient thermo-mechanical analysis of functionally graded hollow circular cylinders *J. Therm. Stresses* **30** 81–104

[11] Abouelregal A E and Tiwari R 2024 Computational analysis of thermoelastic vibrations of functionally graded nonlocal nanobeam excited by thermal shock *J. Vibration and Control* **30** 3105–16

[12] Abouelregal A E and Mohamed B O 2018 Fractional order thermoelasticity for a functionally graded thermoelastic nanobeam induced by a sinusoidal pulse heating *J. Comput. Theor. Nanosci.* **15** 1233–42

[13] Abouelregal A and Tiwari R 2022 The thermoelastic vibration of nano-sized rotating beams with variable thermal properties under axial load via memory-dependent heat conduction *Meccanica* **57** 2001–25

[14] Lotfy K and Sarkar N 2017 Memory-dependent derivatives for photothermal semiconducting medium in generalized thermoelasticity with two-temperature *Mech. Time-Depend. Mater.* **21** 519–34

[15] Ezzat M A, El-Karamany A S and El-Bary A A 2014 Generalized thermo-viscoelasticity with memory-dependent derivatives *Int. J. Mech. Sci.* **89** 470–5

[16] El-Karamany A S and El-Bary A A 2016 Electro-thermoelasticity theory with memory-dependent derivative heat transfer *Int. J. Eng. Sci.* **99** 22–38

[17] Tiwari R, Saeed A M, Abouelregal A, Singhal A and Salem M G 2022 Nonlocal thermoelastic waves inside nanobeam resonator subject to various loadings *Mech. Based Des. Struct. Mach.* **52** 1–24

Chapter 2

Flexural characteristics of functionally graded Timoshenko–Ehrenfest nanobeam

S Ali Faghidian

The mixture unified gradient theory of elasticity is invoked for the nanoscopic study of the flexural characteristics of nanobeams. To appropriately account for the shear deformation, an elastic short stubby nanobeam consistent with the Timoshenko–Ehrenfest beam kinematics is considered. Effective material properties of the nanobeam are assumed to continuously vary across the beam thickness, and accordingly, a functionally graded beam with a general form of the distribution function is considered. A stationary variational framework is established integrating all the governing equations into a single functional. The associated boundary-value problem of equilibrium is determined and enriched with the consistent form of additional nonstandard boundary conditions. To effectively realize the peculiar size-dependent response of nanobeams at the ultra-small scale, the stress gradient theory, the strain gradient theory, and the classical elasticity theory are unified within the framework of the mixture unified gradient theory of elasticity. The flexural characteristics of the mixture unified gradient beam are addressed analytically. A comprehensive numerical study is performed to demonstrate the flexural features of a Timoshenko–Ehrenfest nanobeam wherein the nanoscopic effects corresponding to the gradient characteristic parameters are examined and discussed. A new benchmark for numerical analysis is, therefore, detected based on the presented flexural response of nanobeams and ensuing numerical results.

2.1 Introduction

Nanostructured materials exhibit significant physical characteristics, and accordingly, have found a variety of implications in pioneering nanoscale engineering

doi:10.1088/978-0-7503-6024-1ch2

systems. Advanced design and optimization of the nanosystems necessitate an accurate description of the behavior of media with microstructures. The classical elasticity theory is well-known to be inadequate in capturing the size-effects. Adopting the augmented elasticity theories is, therefore, inevitable [1–3]. Dissimilar forms of the augmented elasticity frameworks are merged in the literature of nanomechanics, yet, there is no consensus on the most precise approach for the nanoscopic study of the field quantities.

Nonlocality and gradient elasticity are, perhaps, the most celebrated size-dependent contexts that are extensively utilized to describe the peculiar behavior of continuum with nanostructural features. The constitutive law of the nonlocal materials is modified in such a way as to incorporate the long-range size-effects. The nonlocal field is, thus, introduced as a weighted mean value in terms of the classical field variables [4]. A different line of thought, departing from the nonlocal mechanics, is the gradient elasticity wherein the constitutive model is characterized by the incorporation of the gradient of field variables [5]. In contrast to the controversial anomalies associated with the nonlocal theory [6], a decisive fact stands in favor of the gradient elasticity theory; it leads to a well-posed boundary-value problem for bounded structural domains [7]. Due to the inherent simplicity associated with nonlocal and gradient elasticity theories, they have been extensively utilized to realize the size-effects in nanoscale structures; to mention just a few representative studies [8–16].

The nonlocal and gradient frameworks can solely capture the smaller-is-softer and the smaller-is-stiffer responses, correspondingly. Depending on particular state conditions of the nanomaterial, both the stiffening and softening structural behaviors are, however, revealed in numerical simulations and experimental measurements. The augmented elasticity theories of the hybrid type are, therefore, proposed in the literature; the instances are the nonlocal strain gradient model [17, 18], the higher-order nonlocal gradient theory [19, 20], the nonlocal surface elasticity [21, 22], the higher-order unified gradient theory [23], and the mixture unified gradient theory [24]. The stationary variational principle associated with the mixture unified gradient theory was recently established in the most general intrinsic form by Faghidian *et al* [24]. The efficacy of the mixture unified gradient theory in nanoscopic analysis of a variety of structural problems in nanomechanics is, subsequently, evinced in a series of very recent publications [25–29].

Within the elasticity framework of the mixture unified gradient theory, the size-effects of the stress gradient theory and the strain gradient theory are integrated while the effects of the classical elasticity theory is, moreover, incorporated. The constitutive law associated with the nanostructured material behavior consistent with the mixture unified gradient theory is, accordingly, enriched with two gradient length-scale parameters along with a mixture parameter. Adopting the conceived variational principle, all the governing equations including the constitutive relations associated with the boundary-value problem of equilibrium are integrated into a single potential functional. The constitutive model of the mixture unified gradient structure is inevitably of high-order in comparison with the classical stress-strain law, and therefore, the suitable form of the additional nonstandard boundary conditions is addressed to close the size-dependent elastic problem on bounded structural domains. Notably, a variety of augmented elasticity theories of the gradient-type can be restored as particular cases of the mixture unified gradient theory of elasticity via adopting ad hoc values of the characteristic length-scale parameters.

This chapter aims to rigorously address the flexural characteristics of a functionally graded Timoshenko–Ehrenfest nanobeam, and it proceeds as follows; to account for the shear deformation in the flexure mechanics of short stubby beams, the first-order shear deformation beam theory is adopted. A proper form of the shear coefficient within the framework of the Timoshenko–Ehrenfest beam model is employed to amend the associated flexural results for the true distribution of shear over the beam cross-section. The beam is considered to be made of functionally graded (FG) material which continuously varies along the thickness direction. The preliminary assumptions on the kinematics of the beam and the functionally graded material behavior are first recalled in section 2.2. A proper form of a variational principle consistent with the Timoshenko–Ehrenfest nanobeam that integrates all the governing equations into a single functional is, afterward, introduced. The differential and boundary conditions of equilibrium are derived based on the stationarity condition of the established variational functional. The associated constitutive model of the Timoshenko–Ehrenfest nanobeam consistent with the mixture unified gradient theory is determined. The boundary-value problem of equilibrium is properly closed with the prescription of the appropriate form of the additional nonstandard boundary conditions. The flexure mechanics of the functionally graded Timoshenko–Ehrenfest nanobeam within the context of the mixture unified gradient theory is analytically addressed in section 2.3. Numerical illustrations of the flexural characteristics of Timoshenko–Ehrenfest nanobeams for boundary conditions of interest in nanomechanics are demonstrated and discussed in section 2.3. The structural softening and stiffening responses of the ultra-small scale Timoshenko–Ehrenfest beam are evinced to be proficiently realized. Concluding remarks are drawn in section 2.4.

2.2 Stationary variational principle

2.2.1 Functionally graded nanobeams

To develop the stationary variational principle, reference is made to a functionally graded straight beam of length $L = b - a$, with a rectangular cross-section $\mathbb{A}$. The beam is referred to as Cartesian orthogonal co-ordinates (x, y, z); the x-axis coincides with the beam longitudinal centroid axis, whereas the z-axis is considered to be parallel to the beam height. The origin of the co-ordinate system is coincident with the cross-sectional geometric center and the plane x–z defines the flexure plane. The beam ends, $x = a, b$, are restrained impeding any global rigid-motion. The beam is, also, assumed to be subjected to a transversal body force per unit length f and an applied distributed flexural couple m. Inertia forces are considered to be negligible. The beam is considered to be made of functionally graded materials which is continuously varied along the thickness direction. The nanobeam material is characterized by the elastic modulus E and the shear modulus G which are assumed to only vary along the z-axis according to a general distribution function

$$E = E(|z|), \quad G = G(|z|) \tag{2.1}$$

The Poisson's ratio ν is considered to be constant in view of ignorable variation through the beam thickness. Without loss of generality, the effective material properties, i.e. elastic and shear moduli, are considered to be symmetric with respect to the centroidal axis. The elastic center and the geometric center of the beam cross-section, thus, coincide. The flexural stiffness I_E is defined by the second moment of elastic area weighted with the scalar field of elastic modulus. Likewise, the shear area A_G is introduced as the elastic cross-sectional area weighted with the scalar field of shear modulus as

$$I_E = \int_{\mathbb{A}} E(|z|) z^2 dA, \quad A_G = \int_{\mathbb{A}} G(|z|) dA \tag{2.2}$$

2.2.2 Mixture unified gradient theory

Within the framework of the Timoshenko–Ehrenfest beam, the centroid axial displacement and the transversal deformation are considered vanishing. As a first-order shear deformation beam model, no warping exists for the beam cross-section after deformation. To set forth the kinematics of the elastic beam, the elasticity solution of Saint–Venant's problem is invoked [30]. The displacement field $\boldsymbol{u}$ consistent with the kinematics of the Timoshenko–Ehrenfest beam theory is written as

$$u_1 = -z\varphi(x), \quad u_2 = 0, \quad u_3 = w(x) \tag{2.3}$$

with φ and w being the cross-sectional rotation and the transverse displacement of the beam. The kinematically compatible strain field, i.e. normal strain ε and the shear strain γ, thus reads as

$$\varepsilon = -z\partial_x\varphi(x) = -z\chi(x), \quad \gamma = -\varphi(x) + \partial_x w(x) \tag{2.4}$$

where $\chi = \partial_x\varphi$ represents the flexural curvature of the centroidal axis of the beam. Based on the intrinsic form of the variational principle associated with the mixture unified gradient theory [24], the potential functional $\mathbb{P}$ consistent with the Timoshenko–Ehrenfest nanobeam is introduced as

$$
\begin{aligned}
\mathbb{P} = \int_a^b \Big[& -M_0(x)\chi(x) + V_0(x)\gamma(x) - M_1(x)\partial_x\chi(x) + V_1(x)\partial_x\gamma(x) \\
& - f(x)w(x) - m(x)\varphi(x) - \frac{1}{2I_E}(M_0(x))^2 - \frac{c^2}{2I_E}(\partial_x M_0(x))^2 \\
& - \frac{1}{2\Bbbk A_G}(V_0(x))^2 - \frac{c^2}{2\Bbbk A_G}(\partial_x V_0(x))^2 - \frac{1}{2I_E(\alpha c^2 + \ell^2)}(M_1(x))^2 \\
& - \frac{c^2}{2I_E(\alpha c^2 + \ell^2)}(\partial_x M_1(x))^2 - \frac{1}{2\Bbbk A_G(\alpha c^2 + \ell^2)}(V_1(x))^2 \\
& - \frac{c^2}{2\Bbbk A_G(\alpha c^2 + \ell^2)}(\partial_x V_1(x))^2 \Big] dx
\end{aligned} \tag{2.5}
$$

for any virtual kinetic fields having compact support on the beam domain. In the introduced functional $\mathbb{P}$, the flexural resultants M_0 and M_1 are, correspondingly, the dual mathematical fields of the flexural curvature χ and of its derivative along the beam longitudinal axis $\partial_x\chi$. Similarly, the shear resultants V_0 and V_1 are introduced as dual mathematical fields of the shear strain γ and of its derivative along the beam axis $\partial_x\gamma$, respectively. The potential functional $\mathbb{P}$ is enriched with the stress gradient characteristic length c and the strain gradient length-scale parameter ℓ to address the significance of the corresponding gradient elasticity theory. The effects of the classical elasticity theory is, moreover, incorporated via the mixture parameter $0 \leqslant \alpha \leqslant 1$. The shear coefficient $\Bbbk$ is adopted within the Timoshenko–Ehrenfest beam framework to take into consideration the non-uniform shear distribution across the beam thickness [31].

Within the context of the stationary variational principles, all the kinematic and kinetic field variables are treated independently of each other. The first variation of the potential functional $\mathbb{P}$, subsequent to integration by parts, is written as

$$
\begin{aligned}
\delta\mathbb{P} = \int_a^b \Big[& -M_0(x)\delta\chi(x) - M_1(x)\partial_x\delta\chi(x) + V_0(x)\delta\gamma(x) \\
& + V_1(x)\partial_x\delta\gamma(x) - f(x)\delta w(x) - m(x)\delta\varphi(x) \\
& - \delta M_0(x)\left(\chi(x) + \frac{1}{I_E}M_0(x) - \frac{c^2}{I_E}\partial_{xx}M_0(x)\right) \\
& - \delta M_1(x)\left(\partial_x\chi(x) + \frac{1}{I_E(\alpha c^2 + \ell^2)}M_1(x) - \frac{c^2}{I_E(\alpha c^2 + \ell^2)}\partial_{xx}M_1(x)\right) \\
& + \delta V_0(x)\left(\gamma(x) - \frac{1}{\Bbbk A_G}V_0(x) + \frac{c^2}{\Bbbk A_G}\partial_{xx}V_0(x)\right) \\
& + \delta V_1(x)\left(\partial_x\gamma(x) - \frac{1}{\Bbbk A_G(\alpha c^2 + \ell^2)}V_1(x) + \frac{c^2}{\Bbbk A_G(\alpha c^2 + \ell^2)}\partial_{xx}V_1(x)\right)\Big] dx \\
& - \frac{c^2}{I_E}\partial_x M_0(x)\delta M_0(x)\Big|_a^b - \frac{c^2}{\Bbbk A_G}\partial_x V_0(x)\delta V_0(x)\Big|_a^b \\
& - \frac{c^2}{I_E(\alpha c^2 + \ell^2)}\partial_x M_1(x)\delta M_1(x)\Big|_a^b - \frac{c^2}{\Bbbk A_G(\alpha c^2 + \ell^2)}\partial_x V_1(x)\delta V_1(x)\Big|_a^b
\end{aligned} \tag{2.6}
$$

The boundary congruence conditions can be released in view of the heuristic assumption on the virtual kinetic test field variables to have compact support on the domain. Adopting the compatibility conditions along with integration by parts with respect to the kinematic field variables, yields

$$
\begin{aligned}
\delta\mathbb{P} = \int_a^b \Bigg[& (\partial_x M_0(x) - \partial_{xx} M_1(x) - V_0(x) + \partial_x V_1(x) - m(x))\delta\varphi(x, t) \\
& - (\partial_x V_0(x) - \partial_{xx} V_1(x) + f(x))\delta w(x) \\
& - \delta M_0(x)\left(\chi(x) + \frac{1}{I_E} M_0(x) - \frac{c^2}{I_E}\partial_{xx} M_0(x) \right) \\
& - \delta M_1(x)\left(\partial_x \chi(x) + \frac{1}{I_E(\alpha c^2 + \ell^2)} M_1(x) - \frac{c^2}{I_E(\alpha c^2 + \ell^2)}\partial_{xx} M_1(x) \right) \\
& + \delta V_0(x)\left(\gamma(x) - \frac{1}{\Bbbk A_G} V_0(x) + \frac{c^2}{\Bbbk A_G}\partial_{xx} V_0(x) \right) \\
& + \delta V_1(x)\left(\partial_x \gamma(x) - \frac{1}{\Bbbk A_G(\alpha c^2 + \ell^2)} V_1(x) + \frac{c^2}{\Bbbk A_G(\alpha c^2 + \ell^2)}\partial_{xx} V_1(x) \right) \Bigg] dx \\
& + (-M_0(x) + \partial_x M_1(x))\delta\varphi|_a^b + (V_0(x) - \partial_x V_1(x))\delta w|_a^b \\
& - M_1(x)\partial_x\delta\varphi|_a^b - V_1(x)\delta(-\varphi + \partial_x w)|_a^b
\end{aligned}
\tag{2.7}
$$

The differential and boundary conditions of equilibrium for a mixture unified gradient Timoshenko–Ehrenfest beam can be achieved via imposing the stationarity condition of the established variational principle, as

$$
\begin{aligned}
& -\partial_x(M_0(x) - \partial_x M_1(x)) + (V_0(x) - \partial_x V_1(x)) + m(x) = 0 \\
& \partial_x(V_0(x) - \partial_x V_1(x)) + f(x) = 0 \\
& (M_0(x) - \partial_x M_1(x))\delta\varphi|_a^b = (V_0(x) - \partial_x V_1(x))\delta w|_a^b = 0 \\
& M_1(x)|_a^b = V_1(x)|_a^b = 0
\end{aligned}
\tag{2.8}
$$

where the flexural curvature and shear strain fields are considered to have arbitrary variations on the beam domain. It is a conventional choice within the context of the gradient elasticity theory to apply the concept of the total flexural moment M and the shear force V as

$$
\begin{aligned}
M(x) &= M_0(x) - \partial_x M_1(x) \\
V(x) &= V_0(x) - \partial_x V_1(x)
\end{aligned}
\tag{2.9}
$$

The differential and boundary conditions of equilibrium can be, thus, cast in the form

$$
\begin{aligned}
& -\partial_x M(x) + V(x) + m(x) = 0 \\
& \partial_x V(x) + f(x) = 0 \\
& M(x)\delta\varphi|_a^b = V(x)\delta w|_a^b = 0 \\
& M_1(x)|_a^b = V_1(x)|_a^b = 0
\end{aligned}
\tag{2.10}
$$

Notably, the boundary-value problem of equilibrium is equipped with the classical boundary conditions along with additional nonstandard boundary conditions, i.e. vanishing of the flexural resultant M_1 and the shear resultant V_1 on the nanobeam ends. The constitutive model associated with the mixture unified gradient Timoshenko–Ehrenfest beam can be, furthermore, determined via prescribing the stationarity condition of the introduced variational functional. The constitutive laws of the resultant moments M_0, M_1, and M are, accordingly, cast as ordinary differential relations expressed by

$$\begin{aligned} M_0(x) - c^2\partial_{xx}M_0(x) &= -I_E\chi(x) \\ M_1(x) - c^2\partial_{xx}M_1(x) &= -I_E(\alpha c^2 + \ell^2)\partial_x\chi(x) \\ M(x) - c^2\partial_{xx}M(x) &= -I_E(\chi(x) - (\alpha c^2 + \ell^2)\partial_{xx}\chi(x)) \end{aligned} \tag{2.11}$$

Likewise, the constitutive laws of the shear resultants V_0, V_1 and the shear force V are determined as

$$\begin{aligned} V_0(x) - c^2\partial_{xx}V_0(x) &= \mathbb{k}A_G\gamma(x) \\ V_1(x) - c^2\partial_{xx}V_1(x) &= \mathbb{k}A_G(\alpha c^2 + \ell^2)\partial_x\gamma(x) \\ V(x) - c^2\partial_{xx}V(x) &= \mathbb{k}A_G(\gamma(x) - (\alpha c^2 + \ell^2)\partial_{xx}\gamma(x)) \end{aligned} \tag{2.12}$$

The constitutive model of the kinetic field variables is enriched with the gradient length-scale characteristic parameters to advantageously comprise the size-effects of the gradient elasticity theory. Enrichment of the constitutive law with the mixture parameter, furthermore, allows one to incorporate the influence of the classical elasticity theory. An efficient augmented elasticity framework is introduced which is capable of a precise description of the nanoscopic response of short stubby elastic nanoscale beams.

The constitutive differential relations of the resultant fields are remarkably of higher-order in comparison with the classical elasticity model. The additional nonstandard boundary conditions should be prescribed to close the mixture unified gradient problem. For accurate analysis of nanoscopic field quantities in beam-type structural problems, explicit mathematical formulae of the flexural resultant M_1 and the shear resultant V_1 should be suitably determined. By following some straightforward mathematics, the flexural and the shear resultants can be determined as

$$\begin{aligned} M_1(x) =& \frac{c^4(\alpha c^2 + \ell^2)}{c^2(1-\alpha) - \ell^2}(-\partial_x f(x) + \partial_{xx}m(x)) \\ &- I_E(\alpha c^2 + \ell^2)\left(\partial_x\chi(x) - \frac{c^2(\alpha c^2 + \ell^2)}{c^2(1-\alpha) - \ell^2}\partial_{xxx}\chi(x)\right) \\ V_1(x) =& \frac{c^4(\alpha c^2 + \ell^2)}{c^2(1-\alpha) - \ell^2}(-\partial_{xx}f(x)) \\ &+ \mathbb{k}A_G(\alpha c^2 + \ell^2)\left(\partial_x\gamma(x) - \frac{c^2(\alpha c^2 + \ell^2)}{c^2(1-\alpha) - \ell^2}\partial_{xxx}\gamma(x)\right) \end{aligned} \tag{2.13}$$

A variety of augmented elasticity theories of the gradient-type can be retrieved as particular cases of the established mixture unified gradient theory of elasticity under ad hoc assumptions on the gradient length-scale characteristic parameters. The classical elasticity model of the Timoshenko–Ehrenfest beam is obtained via either setting the gradient characteristic lengths to zero or via letting the mixture parameter approach unity in the absence of the strain gradient length-scale parameter. Vanishing of the mixture parameter yields the unified gradient elasticity theory. The strain gradient elasticity theory is recovered as the stress gradient length-scale parameter tends to zero. Likewise, the stress gradient theory can be recovered via setting the strain gradient length-scale parameter to zero. The mixture stress gradient theory can be retrieved through vanishing the strain gradient length-scale parameter. A comprehensive discussion on the pertinent size-dependent elasticity theories of gradient-type is addressed by Faghidian and Elishakoff [27].

The peculiar size-dependent flexural characteristics of nanoscale short stubby beams can be competently captured within the framework of the mixture unified gradient theory as evinced via rigorous flexural analysis of Timoshenko–Ehrenfest nanobeams.

2.3 Flexure mechanics of nanobeams

2.3.1 Analytical solution of the flexure

To rigorously study the flexural characteristics of a nanoscale Timoshenko–Ehrenfest beam, the transverse displacement of the short stubby elastic beam consistent with the mixture unified gradient theory is analytically addressed. The common choice usually preferred in mechanics of nanostructures is to describe the differential conditions of equilibrium in terms of kinematics field variables. The resulting higher-order differential equations should be, subsequently, solved. Nevertheless, a variety of solution methodologies exist in the literature to deal with the structural problems wherein the kinematic or the kinetic field variables are assumed to have a series solution form. As a different school of thought, the governing equations can be described in terms of kinetics field variables [32, 33] or the autonomous series solution of both field variables [34, 35]. A viable methodology is invoked here to address the exact analytical solution of the kinematics field variables [36]. The mixture unified gradient Timoshenko–Ehrenfest beam is considered to be subjected to a uniformly distributed transverse load $\overline{f}$ in the absence of the distributed flexural couple. In view of the differential conditions of equilibrium equation (2.10), lines 1 and 2, and the introduced constitutive law equations (2.11) and (2.12)(third line), the flexural moment M and the shear force V are first determined as

$$\begin{aligned} M(x) &= -c^2 f(x) - I_E(\chi(x) - (\alpha c^2 + \ell^2)\partial_{xx}\chi(x)) \\ V(x) &= -c^2 \partial_x f(x) + \Bbbk A_G(\gamma(x) - (\alpha c^2 + \ell^2)\partial_{xx}\gamma(x)) \end{aligned} \tag{2.14}$$

Substitution of the stress resultant fields M and V into the differential conditions of equilibrium, a set of governing equations for the Timoshenko–Ehrenfest nanobeam is achieved as

$$\begin{aligned} &I_E(\partial_x\chi(x) - (\alpha c^2 + \ell^2)\partial_{xxx}\chi(x)) + \Bbbk A_G(\gamma(x) - (\alpha c^2 + \ell^2)\partial_{xx}\gamma(x)) = 0 \\ &f(x) - c^2\partial_{xx}f(x) + \Bbbk A_G(\partial_x\gamma(x) - (\alpha c^2 + \ell^2)\partial_{xxx}\gamma(x)) = 0 \end{aligned} \tag{2.15}$$

The functionally graded material properties of the nanobeam is, solely, assumed in the beam thickness direction, and thus, the flexural stiffness and the shear area are constant along the nanosized beam axis. Solving the ordinary differential equation, line 2 of (2.15) for a uniformly distributed transverse load $\overline{f}$, the analytical solution of the shear strain field γ is determined as

$$\begin{aligned} \gamma(x) = &-\frac{\overline{f}}{\Bbbk A_G}x + \mathbb{C}_1\sqrt{\alpha c^2 + \ell^2}\exp\left(\frac{x}{\sqrt{\alpha c^2 + \ell^2}}\right) \\ &- \mathbb{C}_2\sqrt{\alpha c^2 + \ell^2}\exp\left(-\frac{x}{\sqrt{\alpha c^2 + \ell^2}}\right) + \mathbb{C}_3 \end{aligned} \tag{2.16}$$

Substituting the analytical solution of the shear strain field γ in the ordinary differential equation (2.15) line 1, and solving for the flexural curvature χ, the analytical solution of the rotation field φ reads as

$$\begin{aligned} \varphi(x) = &\frac{\overline{f}}{6I_E}x^3 - \mathbb{C}_3\frac{\Bbbk A_G}{2I_E}x^2 + \mathbb{C}_4(\alpha c^2 + \ell^2)\exp\left(\frac{x}{\sqrt{\alpha c^2 + \ell^2}}\right) \\ &+ \mathbb{C}_5(\alpha c^2 + \ell^2)\exp\left(-\frac{x}{\sqrt{\alpha c^2 + \ell^2}}\right) + \mathbb{C}_6x + \mathbb{C}_7 \end{aligned} \tag{2.17}$$

Lastly, the transverse displacement field w is determined via integrating the kinematic compatibility relation of the shear strain field as

$$\begin{aligned} w(x) = &\frac{\overline{f}}{24I_E}x^4 - \frac{\overline{f}}{2\Bbbk A_G}x^2 + \mathbb{C}_3\left(x - \frac{\Bbbk A_G}{6I_E}x^3\right) \\ &+ (\alpha c^2 + \ell^2)(\mathbb{C}_1 + \mathbb{C}_4\sqrt{\alpha c^2 + \ell^2})\exp\left(\frac{x}{\sqrt{\alpha c^2 + \ell^2}}\right) \\ &+ (\alpha c^2 + \ell^2)(\mathbb{C}_2 - \mathbb{C}_5\sqrt{\alpha c^2 + \ell^2})\exp\left(-\frac{x}{\sqrt{\alpha c^2 + \ell^2}}\right) + \mathbb{C}_6\frac{x^2}{2} + \mathbb{C}_7x + \mathbb{C}_8 \end{aligned} \tag{2.18}$$

To determine the eight unknown integration constants $\mathbb{C}_m(m = 1.8)$, four well-established classical boundary conditions of the Timoshenko–Ehrenfest beam equation (2.10)line 3, along with four additional nonstandard boundary conditions equation (2.13) associated with the mixture unified gradient Timoshenko–Ehrenfest beam should be prescribed. The exact analytical solution of the transverse displacement field of the short stubby nanosized beam consistent with the established mixture unified gradient theory is, therefore, derived.

2.3.2 Numerical results and discussion

To examine the size-effect in the flexural response of short stubby elastic nanobeams, two boundary conditions of interest in the mechanics of nanostructures, i.e. cantilever and fixed-end boundary conditions, are considered. The mixture unified gradient elastic beam is assumed to be subjected to a uniformly distributed transverse load in the absence of the distributed flexural couple. For the Timoshenko–Ehrenfest nanobeam with cantilever boundary conditions, the transverse displacement of the mixture unified gradient elastic beam should meet the classical boundary conditions including the vanishing of the kinematics field variables, i.e. cross-sectional rotation φ and transverse displacement w, at the fixed end along with the vanishing of the kinetic field variables, i.e. the flexural moment M and the shear force V, at the free ends, respectively. Within the framework of the mixture unified gradient theory, the Timoshenko–Ehrenfest nanobeam with cantilever boundary conditions is, furthermore, subjected to the additional nonstandard boundary conditions as the flexural resultant M_1 and the shear resultant V_1 tend to zero at both ends.

Likewise, the essential set of classical boundary conditions associated with the mixture unified gradient elastic beam with fixed-end boundary conditions consists of vanishing the kinematic field variables, i.e. cross-sectional rotation φ and transverse displacement w, at both ends of the beam. The additional nonstandard boundary conditions for the Timoshenko–Ehrenfest nanobeam with fixed-end boundary conditions include setting zero the flexural resultant M_1 and the shear resultant V_1 at the beam ends. Notably, the additional nonstandard boundary conditions associated with the mixture unified gradient elastic beam are independent of the well-recognized classical boundary conditions and they are symmetric with respect to the beam longitudinal axis. The aforementioned set of classical and additional nonstandard boundary conditions consistently closes the boundary-value problem of equilibrium.

For the sake of appropriate comparison, the nondimensional form of the radius of gyration $\hbar$, stress gradient characteristic parameter ζ, strain gradient characteristic parameter η, and transverse deflection of the nanosized beam $\overline{w}$ are introduced as

$$\hbar = \frac{1}{L}\sqrt{\frac{1}{2(1+\nu)}\frac{I_E}{A_G}},\ \zeta = \frac{c}{L},\ \eta = \frac{\ell}{L},\ \overline{w} = w\frac{I_E}{\overline{f}\,L^4} \tag{2.19}$$

Nanoscopic effects of the gradient length-scale parameters on the flexural response of the mixture unified gradient Timoshenko–Ehrenfest beam are demonstrated in figures 2.1 and 2.2 for cantilever and fixed-ends boundary conditions, respectively. In the ensuing numerical results, the gradient characteristic parameters range in the interval $[0, 1/2]$ while three values of the mixture parameter $\alpha = 0, 1/5, 1$ are correspondingly prescribed. Assuming the Poisson's ratio as

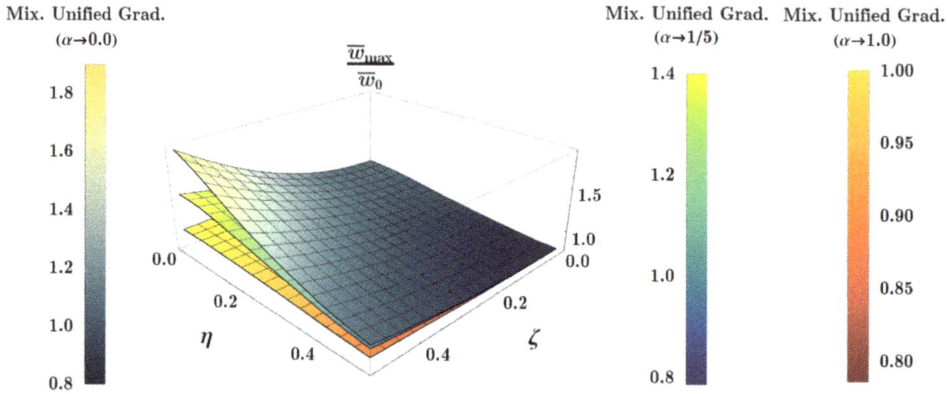

Figure 2.1. Effects of characteristic parameters on the flexural response of a uniformly loaded cantilever Timoshenko–Ehrenfest nanobeam.

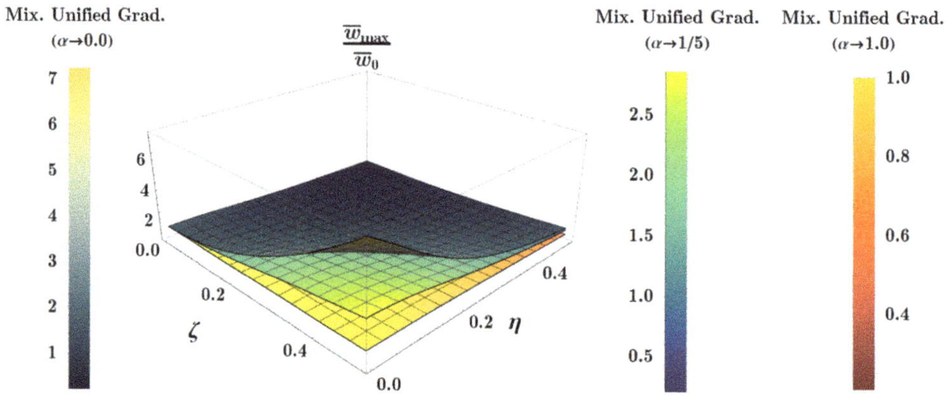

Figure 2.2. Effects of characteristic parameters on the flexural response of a uniformly loaded fixed-end Timoshenko–Ehrenfest nanobeam.

$\nu = 1/5$, the shear coefficient is adopted for the rectangular cross-section [31]. The nondimensional radius of gyration is prescribed as $\hbar = 1/10$. The maximum transverse displacement of the nanosized Timoshenko–Ehrenfest beam is, also, normalized with respect to the pertinent flexural response of the classical Timoshenko–Ehrenfest beam model $\overline{w}_0$, for illustrative purposes.

It is notably inferred from the demonstrated numerical results associated with the flexural response of the Timoshenko–Ehrenfest nanobeam that the normalized transverse displacement of the mixture unified gradient beam is decreased by increasing the strain gradient characteristic parameter η, i.e. a larger value of η involves a smaller normalized deflection for prescribed stress gradient characteristic parameter and the mixture parameter. A stiffening response in terms of the strain gradient characteristic parameter is, therefore, realized within the framework of the mixture unified gradient theory. In contrast, the stress gradient characteristic parameter ζ has the effect of increasing the transverse displacement of the nanosized beam, i.e. a larger value of ζ involves a larger deflection for prescribed strain gradient and mixture parameters. A softening response in terms of the stress gradient characteristic parameter is, accordingly, confirmed for the mixture unified gradient elasticity framework. Nanoscopic effects of the stress gradient theory are reduced and reinstated with the effects of the classical elasticity theory as the mixture parameter α continuously varies from zero to unity. For prescribed values of the gradient characteristic parameters, the normalized transverse displacement of the nanobeam is realized to be decreased with increasing the mixture parameter α. A stiffening response in terms of the mixture parameter α is captured for the mixture unified gradient theory. The peculiar size-dependent response of the short stubby nanosized beam can be advantageously realized within the context of the mixture unified gradient theory. The classical flexural response of the Timoshenko–Ehrenfest beam is inevitably retrieved as either the gradient characteristic parameters approach zero or in the absence of the strain gradient characteristic parameter while the mixture parameter tends to unity. As expected, a stiffening response in terms of the number of kinematic boundary constraints is realized within the augmented elasticity framework of the mixture unified gradient theory. A careful examination of the numerical illustrations, also, reveals that the flexural response of the mixture unified gradient Timoshenko–Ehrenfest beam is less affected by the gradient characteristic parameters for non-vanishing values of the mixture parameter.

Tables 2.1 and 2.2, respectively, summarize the normalized maximum transverse displacement of the nanosized Timoshenko–Ehrenfest beam, detected within the size-dependent elasticity framework of the mixture unified gradient theory for cantilever and fixed-end boundary conditions adopting different values of the mixture parameter. Numerical results of the normalized transverse displacement of the Timoshenko–Ehrenfest nanobeam with $\alpha = 1/2$ are not demonstrated in the numerical illustrations, for the sake of clarity, but are tabulated in tables 2.1 and 2.2.

2.4 Concluding remarks

This study aims to shed light on the flexural characteristics of the nanoscale functionally graded Timoshenko–Ehrenfest beam. The mixture unified gradient theory is invoked for a rigorous nanoscopic analysis of the short stubby elastic beams. The stress gradient theory and the strain gradient theory are consistently unified to realize the peculiar size-effects at the ultra-small scale while the structural effect of the classical elasticity theory is appropriately incorporated too. The material properties of the nanobeam are assumed to continuously vary along the beam thickness direction with a general form of the distribution function. Without loss of generality, the effective material properties are considered to be symmetric with respect to the centroidal axis. The elastic center and the geometric center of the beam cross-section, therefore, coincide. The pertinent mathematical forms of the flexural stiffness are introduced based on the effective moduli of the functionally graded material. The kinematics of the Timoshenko–Ehrenfest beam model is

Table 2.1. Normalized maximum deflection of a uniformly loaded cantilever Timoshenko–Ehrenfest nanobeam.

$\overline{w}_{max}/\overline{w}_0$												
	$\alpha \to 0$						$\alpha \to 1/5$					
ζ	$\eta = 0$	$\eta = 0.1$	$\eta = 0.2$	$\eta = 0.3$	$\eta = 0.4$	$\eta = 0.5$	$\eta = 0$	$\eta = 0.1$	$\eta = 0.2$	$\eta = 0.3$	$\eta = 0.4$	$\eta = 0.5$
0	1.000 00	0.971 24	0.912 96	0.857 23	0.815 17	0.785 76	1.000 00	0.971 24	0.912 96	0.857 23	0.815 17	0.785 76
0.1	1.035 95	1.000 00	0.934 72	0.873 09	0.826 72	0.794 33	1.026 18	0.994 39	0.931 41	0.871 30	0.825 76	0.793 80
0.2	1.143 78	1.086 27	1.000 00	0.920 68	0.861 38	0.820 04	1.094 45	1.057 89	0.983 58	0.911 77	0.856 58	0.817 37
0.3	1.323 51	1.230 06	1.108 80	1.000 00	0.919 14	0.862 89	1.189 44	1.148 66	1.061 34	0.973 95	0.904 97	0.854 94
0.4	1.575 12	1.431 35	1.261 12	1.111 05	1.000 00	0.922 87	1.296 71	1.252 93	1.154 54	1.051 48	0.967 11	0.904 23
0.5	1.898 63	1.690 17	1.456 95	1.253 82	1.103 97	1.000 00	1.404 66	1.359 48	1.253 82	1.137 74	1.038 72	0.962 52

$\overline{w}_{max}/\overline{w}_0$												
	$\alpha \to 1/2$						$\alpha \to 1$					
ζ	$\eta = 0$	$\eta = 0.1$	$\eta = 0.2$	$\eta = 0.3$	$\eta = 0.4$	$\eta = 0.5$	$\eta = 0$	$\eta = 0.1$	$\eta = 0.2$	$\eta = 0.3$	$\eta = 0.4$	$\eta = 0.5$
0	1.000 00	0.971 24	0.912 96	0.857 23	0.815 17	0.785 76	1.000 00	0.971 24	0.912 96	0.857 23	0.815 17	0.785 76
0.1	1.015 43	0.986 43	0.926 62	0.868 67	0.824 35	0.793 01	1.000 00	0.974 20	0.919 07	0.864 48	0.822 06	0.791 73
0.2	1.051 59	1.023 57	0.962 16	0.899 68	0.849 89	0.813 58	1.000 00	0.979 77	0.932 95	0.882 36	0.839 94	0.807 79
0.3	1.094 35	1.068 44	1.008 15	0.942 44	0.886 85	0.844 39	1.000 00	0.984 94	0.947 71	0.903 50	0.862 89	0.829 67
0.4	1.134 11	1.111 05	1.054 69	0.988 88	0.929 42	0.881 48	1.000 00	0.988 88	0.959 99	0.922 87	0.885 79	0.853 08
0.5	1.167 06	1.147 10	1.096 33	1.033 34	0.972 74	0.921 12	1.000 00	0.991 67	0.969 25	0.938 68	0.905 97	0.875 14

Table 2.2. Normalized maximum deflection of a uniformly loaded fixed-end Timoshenko–Ehrenfest nanobeam.

$\overline{w}_{max}/\overline{w}_0$												
	$\alpha \to 0$						$\alpha \to 1/5$					
ζ	$\eta = 0$	$\eta = 0.1$	$\eta = 0.2$	$\eta = 0.3$	$\eta = 0.4$	$\eta = 0.5$	$\eta = 0$	$\eta = 0.1$	$\eta = 0.2$	$\eta = 0.3$	$\eta = 0.4$	$\eta = 0.5$
0	1.000 00	0.831 15	0.583 81	0.404 20	0.286 12	0.208 88	1.000 00	0.831 15	0.583 81	0.404 20	0.286 12	0.208 88
0.1	1.249 95	1.000 00	0.687 86	0.470 40	0.330 73	0.240 53	1.170 78	0.967 71	0.674 82	0.464 70	0.327 98	0.239 08
0.2	1.999 81	1.506 54	1.000 00	0.669 00	0.464 59	0.335 46	1.567 00	1.315 56	0.923 97	0.636 48	0.449 12	0.327 42
0.3	3.249 56	2.350 78	1.520 24	1.000 00	0.687 68	0.493 68	2.032 74	1.755 90	1.274 92	0.894 18	0.637 29	0.467 48
0.4	4.999 23	3.532 71	2.248 58	1.463 40	1.000 00	0.715 20	2.474 93	2.199 12	1.668 06	1.205 67	0.875 46	0.649 86
0.5	7.248 79	5.052 33	3.185 02	2.059 20	1.401 56	1.000 00	2.860 87	2.601 95	2.059 20	1.540 61	1.145 28	0.863 63

$\overline{w}_{max}/\overline{w}_0$												
	$\alpha \to 1/2$						$\alpha \to 1$					
ζ	$\eta = 0$	$\eta = 0.1$	$\eta = 0.2$	$\eta = 0.3$	$\eta = 0.4$	$\eta = 0.5$	$\eta = 0$	$\eta = 0.1$	$\eta = 0.2$	$\eta = 0.3$	$\eta = 0.4$	$\eta = 0.5$
0	1.000 00	0.831 15	0.583 81	0.404 20	0.286 12	0.208 88	1.000 00	0.831 15	0.583 81	0.404 20	0.286 12	0.208 88
0.1	1.096 14	0.924 10	0.656 28	0.456 42	0.323 94	0.236 95	1.000 00	0.861 56	0.627 83	0.443 29	0.317 43	0.233 49
0.2	1.276 88	1.118 47	0.831 37	0.593 49	0.427 80	0.316 06	1.000 00	0.906 96	0.715 02	0.533 91	0.396 51	0.298 80
0.3	1.441 92	1.309 55	1.034 31	0.773 06	0.574 44	0.433 06	1.000 00	0.938 14	0.792 85	0.632 16	0.493 68	0.385 80
0.4	1.569 96	1.463 40	1.218 97	0.957 34	0.738 44	0.571 73	1.000 00	0.957 34	0.849 13	0.715 20	0.586 40	0.476 51
0.5	1.665 29	1.579 96	1.370 74	1.124 81	0.900 42	0.717 40	1.000 00	0.969 34	0.887 81	0.778 89	0.664 96	0.559 89

adopted to properly take into consideration the shear deformation of the short stubby beams. The stationary functional associated with the Timoshenko–Ehrenfest nanobeam is introduced within the mixture unified gradient elasticity framework. All the governing equations are, therefore, integrated into a single functional. The boundary-value problem of equilibrium is established and properly equipped with additional nonstandard boundary conditions. The explicit mathematical formulae of the flexural and shear resultants are determined as well, to be prescribed in the additional nonstandard boundary conditions. A range of augmented elasticity theories of the gradient-type are demonstrated to be recovered as special cases of the established size-dependent elasticity framework under ad hoc assumptions on the gradient length-scale characteristic parameters. The flexure mechanics of the Timoshenko–Ehrenfest nanobeam is meticulously formulated. A viable analytical approach is addressed to determine the exact closed-form solution of the kinematics field variables of the mixture unified gradient beam, i.e. the transverse displacement and rotation field of the short stubby nanosized beam. The numerical results associated with the flexure of the Timoshenko–Ehrenfest nanobeam for boundary conditions of interest in nanomechanics, i.e. cantilever and fixed-end boundary conditions, are demonstrated and discussed. Nanoscopic effects of the gradient length-scale parameters on the flexural features of the mixture unified gradient Timoshenko–Ehrenfest beam are illustrated. The structural softening and stiffening responses of the ultra-small scale Timoshenko–Ehrenfest beam are evinced to be proficiently realized. A new benchmark for numerical analysis is detected based on the presented flexural characteristics of Timoshenko–Ehrenfest nanobeams. The introduced rigorous nanoscopic analysis can be advantageously employed in the design and optimization of beam-type elements of advanced nanoscale engineering systems.

Funding

This research did not receive any specific grant from funding agencies in the public, commercial, or not-for-profit sectors.

Declaration of competing interest

The author declares that he has no known competing financial interests or personal relationships that could have appeared to influence the work reported here.

References

[1] Zhang J and Hoshino K 2018 *Molecular Sensors and Nanodevices, Principles, Designs and Applications in Biomedical Engineering* (Amsterdam: Elsevier)

[2] Elishakoff I, Pentaras D and Gentilini C 2016 *Functionally Graded Material Structures* (Hackensack, NJ: World Scientific)

[3] Elishakoff I *et al* 2012 *Carbon Nanotubes and Nano Sensors: Vibrations, Buckling, and Ballistic Impact* (London: ISTE-Wiley)

[4] Eringen A C 2002 *Nonlocal Continuum Field Theories* (New York: Springer)

[5] Polizzotto C 2015 A unifying variational framework for stress gradient and strain gradient elasticity theories *Eur. J. Mech.* A **49** 430–40

[6] Romano G and Diaco M 2021 On formulation of nonlocal elasticity problems *Meccanica* **56** 1303–28

[7] Polizzotto C 2014 Stress gradient versus strain gradient constitutive models within elasticity *Int. J. Solids Struct.* **51** 1809–18

[8] Ding W, Patnaik S and Semperlotti F 2022 Multiscale nonlocal elasticity: a distributed order fractional formulation *Int. J. Mech. Sci.* **226** 107381

[9] Patnaik S, Sidhardh S and Semperlotti F 2022 Displacement-driven approach to nonlocal elasticity *Eur. J. Mech.* A **92** 104434

[10] Singh B 2022 Propagation characteristics of plane waves in nonlocal isotropic diffusive materials *Appl. Math. Modelling* **104** 306–14

[11] Darban H, Luciano R and Basista M 2022 Calibration of the length scale parameter for the stress-driven nonlocal elasticity model from quasistatic and dynamic experiments *Mech. Adv. Mater. Struct.*

[12] Kaur B and Singh B 2021 Rayleigh-type surface wave in nonlocal isotropic diffusive materials *Acta Mech.* **232** 3407–16

[13] Hache F, Challamel N and Elishakoff I 2019 Asymptotic derivation of nonlocal plate models from three-dimensional stress gradient elasticity *Contin. Mech. Thermodyn.* **31** 47–70

[14] Hache F, Challamel N and Elishakoff I 2019 Asymptotic derivation of nonlocal beam models from two dimensional nonlocal elasticity *Math. Mech. Solids* **24** 2425–43

[15] Jena S K, Chakraverty S and Tornabene F 2019 Buckling behavior of nanobeams placed in electromagnetic field using shifted Chebyshev polynomials-based Rayleigh–Ritz method *Nanomaterials* **9** 1326

[16] Jena S K, Chakraverty S and Tornabene F 2019 Dynamical behavior of nanobeam embedded in constant, linear, parabolic, and sinusoidal types of Winkler elastic foundation using first-order nonlocal strain gradient model *Mater. Res. Express* **6** 0850f2

[17] Aifantis A C 2003 Update on a class of gradient theories *Mech. Mater.* **35** 259–80

[18] Aifantis A C 2011 On the gradient approach—relation to Eringen's nonlocal theory *Int. J. Eng. Sci.* **49** 1367–77

[19] Żur K K and Faghidian S A 2024 Nanomechanics of Structures and Materials: Modeling and Analysis (Amsterdam: Elsevier) https://doi.org/10.1016/C2023-0-00141-8

[20] Faghidian S A 2020 Higher-order nonlocal gradient elasticity: a consistent variational theory *Int. J. Eng. Sci.* **154** 103337

[21] Li L, Lin R and Ng T Y 2020 Contribution of nonlocality to surface elasticity *Int. J. Eng. Sci.* **152** 103311

[22] Jiang Y, Li L and Hu Y 2022 A nonlocal surface theory for surface–bulk interactions and its application to mechanics of nanobeams *Int. J. Eng. Sci.* **172** 103624

[23] Faghidian S A, Zur K K and Reddy J N 2022 A mixed variational framework for higher-order unified gradient elasticity *Int. J. Eng. Sci.* **170** 103603

[24] Faghidian S A, Żur K K and Pan E 2023 Stationary variational principle of mixture unified gradient elasticity *Int. J. Eng. Sci.* **182** 103786

[25] Faghidian S A and Elishakoff I 2023 A consistent approach to characterize random vibrations of nanobeams *Eng. Anal. Bound. Elem.* **152** 14–21

[26] Faghidian S A, Żur K K and Elishakoff I 2023 Nonlinear flexure mechanics of mixture unified gradient nanobeams *Commun. Nonlinear Sci. Numer. Simul.* **117** 106928

[27] Faghidian S A and Elishakoff I 2022 Wave propagation in Timoshenko–Ehrenfest nanobeam: a mixture unified gradient theory *ASME J. Vib. Acoust.* **144** 061005
[28] Faghidian S A and Tounsi A 2022 Dynamic characteristics of mixture unified gradient elastic nanobeams *Facta Univ. Ser.: Mech. Eng.* **20** 539–52
[29] Faghidian S A, Żur K K and Rabczuk T 2022 Mixture unified gradient theory: a consistent approach for mechanics of nanobars *Appl. Phys.* A **128** 996
[30] Faghidian S A 2016 Unified formulation of the stress field of Saint–Venant's flexure problem for symmetric cross-sections *Int. J. Mech. Sci.* 111–2 65–72
[31] Faghidian S A and Elishakoff I 2023 The tale of shear coefficients in Timoshenko–Ehrenfest beam theory: 130 years of progress *Meccanica* **58** 97–108
[32] Faghidian S A 2017 Analytical inverse solution of eigenstrains and residual fields in autofrettaged thick-walled tubes *ASME J. Pressure Vessel Technol.* **139** 031205
[33] Faghidian S A 2017 Analytical approach for inverse reconstruction of eigenstrains and residual stresses in autofrettaged spherical pressure vessels *ASME J. Pressure Vessel Technol.* **139** 041202
[34] Żur K K and Faghidian S A 2021 Analytical and meshless numerical approaches to unified gradient elasticity theory *Eng. Anal. Bound. Elem.* **130** 238–48
[35] Faghidian S A, Żur K K, Pan E and Kim J 2022 On the analytical and meshless numerical approaches to mixture stress gradient functionally graded nano-bar in tension *Eng. Anal. Bound. Elem.* **134** 571–80
[36] Faghidian S A 2021 Flexure mechanics of nonlocal modified gradient nanobeams *J. Comput. Des. Eng.* **8** 949–59

IOP Publishing

Advances in Modeling and Analysis of Functionally Graded Micro- and Nanostructures

Subrat Kumar Jena, S Pradyumna and S Chakraverty

Chapter 3

Free vibration of functionally graded strain gradient nanobeams restrained with elastic springs

Büşra Uzun and Mustafa Özgür Yaylı

Functionally graded composites have various advantages by exhibiting superior properties than the different materials that form them. In addition, it is known that a material exhibits superior properties compared to its macro counterparts at the nanoscale, and when the properties of functionally graded materials are combined with small scales, unique materials emerge. The subject of this study is the size-dependent vibration analysis of functionally graded nanobeams restrained by elastic springs. The size-dependency is considered by the strain gradient theory. At the end of the study, an eigenvalue problem is obtained that can calculate the free vibrational frequencies of functionally graded strain gradient nanobeams with both rigid and elastic boundary conditions. This eigenvalue problem of the functionally graded nanobeam can be used for arbitrary boundary condition and does not require re-solution of the problem as the boundary condition changes. The eigenvalue problem obtained by applying Fourier sine series and Stokes' transform together includes elastic spring stiffnesses, size parameter, and functionally graded strain gradient nanobeam properties that vary depending on the material grading coefficient. It should also be noted that assigning very small and very large values to the spring stiffnesses causes these springs behave like rigid boundaries, which have been widely studied in the literature.

3.1 Introduction

Functionally graded (FG) composite materials consist of two or more components with different properties and, as the name suggests, the transition between the components occurs in a graded manner. The graded transition of properties gives

doi:10.1088/978-0-7503-6024-1ch3

these materials various advantages. The continuous transition in functionally graded composite materials can reduce the residual stress level by a significant amount, leading to improved mechanical and physical properties [1]. In addition, it has also been found that structural quality and ductility are maintained when using these materials. These important properties of functionally graded composite materials have helped in the development of today's applications such as biomedical devices/ prostheses and solar panels [2]. Functionally graded composite materials have been used by many researchers to design and analyze structural elements. The authors in [3–6] presented vibration analysis of beams made of FG material. The authors in [7, 8] studied the static of FG beams. The authors in [9–12] studied both vibration and bending of beams with FG material.

FG materials have also attracted the attention of scholars studying structural elements at the nano/micro scale. The reduction in size adds new properties to the materials, leading to unusual behavior. Various higher-order elasticity theories [13–20] have been proposed to study these behaviors faster and with less cost than experiments and molecular dynamics simulations. The difference of these proposed theories is that they include scale parameters that are not included in the classical theory. Some higher-order elasticity theories include one scale parameter and some include more than one scale parameter and can analyze size effects that cannot be analyzed by classical theories. Papargyri-Beskou *et al* [21] have shown the static and buckling of thin elastic beams via a simple strain gradient elasticity theory (SGET). The initial values approach has been used by Artan and co-workers [22, 23] to calculate the frequencies and buckling loads of beams with different end conditions and via SGET. Ramezani [24] has presented a geometrically nonlinear thick microbeam model in the context of SGET. Yayli [25] has proposed the finite element model of gradient elastic beams for stability response. Gül [26] has investigated the buckling of short-fiber reinforced composite nanobeams based on SGET.

In this chapter, we present a solution approach for the free dynamic of a nanobeam composed of FG materials under deformable boundary conditions. The functionally graded nanobeam is modeled with lateral springs at both ends. Using a methodology based on Fourier sine series and Stokes' transformation, an eigenvalue problem involving the material characteristic length and the stiffnesses of the deformable lateral springs is constructed. The solution of this general problem can give both the results of the deformable boundary conditions and the results of the rigid boundary conditions with an appropriate adjustment of the spring stiffnesses. This solution, which includes functionally graded material properties, is able to calculate the frequencies of both free and simply-supported nanobeams and the frequencies of deformable supported nanobeams between these two boundary conditions, as well as the results of the classical theory by neglecting the size parameter.

3.2 Material properties of functionally graded composite

In this work, the vibration of a FG nanobeam along the height and constrained by lateral springs will be presented by strain gradient theory. The functionally graded

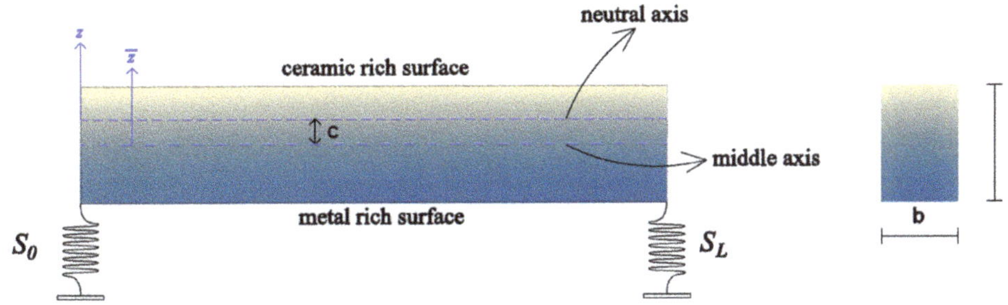

Figure 3.1. Functionally graded nanobeam restrained with lateral springs.

nanobeam is full ceramic on the top surface and full metal on the bottom surface. The nanobeam to be analyzed has width b, height h, and length L. A height-dependent function will be used to calculate the modulus of elasticity (E) and mass density (ρ) of the functionally graded nanobeam. This function is called the power-law distribution and, as can be seen from the literature, it has been adopted and widely used by researchers. The variations of the modulus of elasticity and mass density with respect to the power-law distribution are expressed as follows [27]:

$$E(\bar{z}) = (E_1 - E_2)\left(\frac{\bar{z}}{h} + \frac{1}{2}\right)^p + E_2 \tag{3.1}$$

$$\rho(\bar{z}) = (\rho_1 - \rho_2)\left(\frac{\bar{z}}{h} + \frac{1}{2}\right)^p + \rho_2 \tag{3.2}$$

In the above expressions, E_1 and ρ_1 are the properties of the ceramic component, while E_2 and ρ_2 are the properties of the metal component. $\bar{z}$ is the height direction based on the middle axis of the nanobeam, as can be seen from figure 3.1. Finally, p is the power-law index. In this study, the power-law distribution will be used together with the neutral axis set. It is known that in composite materials the neutral axis is often not the same as the geometric axis. To find the neutral axis position, equation (3.3) is defined [27]:

$$C = \frac{\int_{-h/2}^{+h/2} E(\bar{z})\bar{z}d\bar{z}}{\int_{-h/2}^{+h/2} E(\bar{z})d\bar{z}} \tag{3.3}$$

3.3 Strain gradient-dependent vibration

In this section, the vibration problem considered with SGET will be presented and then the solution method will be applied. For this purpose, this section is considered with two subsections. In section 3.3.1, the governing equation and boundary conditions given in the work presented by [22] will be adapted to the beam model consisting of FG material. In the second subsection (section 3.3.2), the eigenvalue problem obtained by the solution method based on Fourier sine series and Stokes' transformation will be presented.

3.3.1 Governing equation

Artan and Batra [22] have investigated the vibrations of beams made of an elastic, isotropic, and homogeneous material. In their study, they have utilized Hamilton's principle to derive the governing equation and boundary conditions and obtained the Hamiltonian H as follows:

$$H = \int_0^{t_1} dt \int_0^L \frac{1}{2}\left[\rho A\left(\frac{\partial v\,(x,\,t)}{\partial t}\right)^2 - EI\left(\frac{\partial^2 v\,(x,\,t)}{\partial x^2}\right)^2 - EIg^2\left(\frac{\partial^3 v\,(x,\,t)}{\partial x^3}\right)^2\right]dx \tag{3.4}$$

Here, A is the cross-sectional area of the beam I is the moment of inertia, t represents time, $v\,(x,\,t)$ is the lateral displacement, and g is the material characteristic length. Starting from the expression obtained above for homogeneous material, the relation giving the vibration of a FG strain gradient nanobeam is re-written as follows:

$$m\frac{\partial^2 v\,(x,\,t)}{\partial t^2} + D_z\frac{\partial^4 v\,(x,\,t)}{\partial x^4} - D_z g^2\frac{\partial^6 v\,(x,\,t)}{\partial x^6} = 0 \tag{3.5}$$

where, m is the mass per unit length of the functionally graded nanobeam and D_z is the bending stiffness. m and D_z are calculated with respect to the neutral axis z as follows:

$$m = \int_{-\frac{h}{2}-C}^{+\frac{h}{2}-C}\left((\rho_1 - \rho_2)\left(\frac{z + C}{h} + \frac{1}{2}\right)^p + \rho_2\right)dz \tag{3.6}$$

$$D_z = \int_{-\frac{h}{2}-C}^{+\frac{h}{2}-C}\left((E_1 - E_2)\left(\frac{z + C}{\mathrm{h}} + \frac{1}{2}\right)^p + E_2\right)z^2 dz \tag{3.7}$$

3.3.2 Eigenvalue problem

In this subsection, the eigenvalue problem to find the frequencies of a functionally graded strain gradient nanobeam will be obtained. To present the lateral dynamic of a functionally graded nanobeam, $v\,(x,\,t)$ is described as follows:

$$v\,(x,\,t) = W(x)e^{i\omega t} \tag{3.8}$$

in which, $W(x)$ is the modal displacement function and ω is the circular frequency. With the help of equation (3.8), equation (3.5) is re-written as follows:

$$-m\omega^2 W(x) + D_z\frac{d^4 W(x)}{dx^4} - D_z g^2\frac{d^6 W(x)}{dx^6} = 0 \tag{3.9}$$

In this study, $W(x)$ is defined in three different expressions, two for $x = 0$ and $x = L$ points and the other for the intermediate region between $x = 0$ and $x = L$ points of the nanobeam as follows [28]:

$$W(x) = \sum_{n=1}^{\infty} B_n \sin\left(\frac{n\pi x}{L}\right) \text{ for } 0 < x < L \tag{3.10}$$

$$W(x) = W_0 \text{ for } x = 0 \tag{3.11}$$

$$W(x) = W_L \text{ for } x = L \tag{3.12}$$

in which, B_n defines the Fourier coefficient. The derivatives of $W(x)$ become with the help of Stokes' transform [28, 29]:

$$\frac{dW(x)}{dx} = \frac{W_L - W_0}{L} + \sum_{n=1}^{\infty}\left(\frac{2((-1)^n W_L - W_0)}{L} + \frac{n\pi}{L} B_n\right)\cos\left(\frac{n\pi x}{L}\right) \tag{3.13}$$

$$\frac{d^2W(x)}{dx^2} = -\sum_{n=1}^{\infty}\frac{n\pi}{L}\left(\frac{2((-1)^n W_L - W_0)}{L} + \frac{n\pi}{L} B_n\right)\sin\left(\frac{n\pi x}{L}\right) \tag{3.14}$$

$$\begin{aligned}\frac{d^3W(x)}{dx^3} = {} & \frac{W_L{}'' - W_0{}''}{L} \\ & + \sum_{n=1}^{\infty}\left(\frac{2((-1)^n W_L{}'' - W_0{}'')}{L}\right. \\ & \left. - \frac{n^2\pi^2}{L^2}\left(\frac{2((-1)^n W_L - W_0)}{L} + \frac{n\pi}{L} B_n\right)\right)\cos\left(\frac{n\pi x}{L}\right)\end{aligned} \tag{3.15}$$

$$\begin{aligned}\frac{d^4W(x)}{dx^4} = {} & -\sum_{n=1}^{\infty}\frac{n\pi}{L}\left(\frac{2((-1)^n W_L{}'' - W_0{}''')}{L}\right. \\ & \left. - \frac{n^2\pi^2}{L^2}\left(\frac{2((-1)^n W_L - W_0)}{L} + \frac{n\pi}{L} B_n\right)\right)\sin\left(\frac{n\pi x}{L}\right)\end{aligned} \tag{3.16}$$

$$\begin{aligned}\frac{d^5W(x)}{dx^5} = {} & \frac{W_L^{(4)} - W_0^{(4)}}{L} \\ & + \sum_{n=1}^{\infty}\left(\frac{2\left((-1)^n W_L^{(4)} - W_0^{(4)}\right)}{L}\right. \\ & - \frac{n^2\pi^2}{L^2}\left(\frac{2((-1)^n W_L{}'' - W_0{}'')}{L}\right) \\ & \left. + \frac{n^4\pi^4}{L^4}\left(\frac{2((-1)^n W_L - W_0)}{L} + \frac{n\pi}{L} B_n\right)\right)\cos\left(\frac{n\pi x}{L}\right)\end{aligned} \tag{3.17}$$

$$\frac{d^6W(x)}{dx^6} = -\sum_{n=1}^{\infty}\frac{n\pi}{L}\left(\frac{2\left((-1)^n W_L^{(4)} - W_0^{(4)}\right)}{L} - \frac{n^2\pi^2}{L^2}\left(\frac{2((-1)^n W_L'' - W_0'')}{L}\right) + \frac{n^4\pi^4}{L^4}\left(\frac{2((-1)^n W_L - W_0)}{L} + \frac{n\pi}{L}B_n\right)\right)\sin\left(\frac{n\pi x}{L}\right) \tag{3.18}$$

When the derivative expressions given above and $W(x)$ are substituted into equation (3.9) and the solution is performed, the Fourier coefficient is calculated as follows:

$$B_n = \frac{2D_z n^3\pi^3(L^2 + g^2n^2\pi^2)(W_0 - (-1)^n W_L)}{D_z n^4\pi^4(L^2 + g^2n^2\pi^2) - L^6 \mathrm{m}\omega^2} \tag{3.19}$$

It should be noted that the condition $W_L'' = W_0'' = W_L^{(4)} = W_0^{(4)} = 0$ is applied when calculating the Fourier coefficient. The force boundary conditions written depending on the shear force of the strain gradient functionally graded nanobeam are as follows:

$$D_z g^2\frac{\partial^4 v\,(x,\,t)}{\partial x^4} - D_z\frac{\partial^2 v\,(x,\,t)}{\partial x^2} = S_0 W_0 \;\text{ for } x = 0 \tag{3.20}$$

$$D_z g^2\frac{\partial^4 v\,(x,\,t)}{\partial x^4} - D_z\frac{\partial^2 v\,(x,\,t)}{\partial x^2} = -S_L W_L \text{ for } x = L \tag{3.21}$$

S_0 and S_L in equations (3.20) and (3.21) represent the stiffnesses of the lateral springs at $x = 0$ and $x = L$, respectively. With the help of the above force boundary conditions, two homogeneous systems of equations are derived:

$$\left(-S_O + \sum_{n=1}^{\infty}\frac{2D_z Ln^2\pi^2(L^2 + g^2n^2\pi^2)m\omega^2}{D_z n^4\pi^4(L^2 + g^2n^2\pi^2) - L^6 m\omega^2}\right)W_0 + \left(-\sum_{n=1}^{\infty}\frac{2(-1)^n D_z Ln^2\pi^2(L^2 + g^2n^2\pi^2)m\omega^2}{D_z n^4\pi^4(L^2 + g^2n^2\pi^2) - L^6 m\omega^2}\right)W_L = 0 \tag{3.22}$$

$$\left(-\sum_{n=1}^{\infty}\frac{2(-1)^n D_z Ln^2\pi^2(L^2 + g^2n^2\pi^2)m\omega^2}{D_z n^4\pi^4(L^2 + g^2n^2\pi^2) - L^6 m\omega^2}\right)W_0 + \left(-S_L + \sum_{n=1}^{\infty}\frac{2D_z Ln^2\pi^2(L^2 + g^2n^2\pi^2)m\omega^2}{D_z n^4\pi^4(L^2 + g^2n^2\pi^2) - L^6 m\omega^2}\right)W_L = 0 \tag{3.23}$$

The above two sets of equations can be re-written as an eigenvalue problem as follows:

$$\begin{bmatrix} \Gamma_{11} & \Gamma_{12} \\ \Gamma_{21} & \Gamma_{22} \end{bmatrix} \begin{bmatrix} W_0 \\ W_L \end{bmatrix} = 0 \tag{3.24}$$

To calculate the frequencies of the functionally graded strain gradient nanobeam, the eigenvalues are calculated by solving equation (3.25):

$$\begin{vmatrix} \Gamma_{11} & \Gamma_{12} \\ \Gamma_{21} & \Gamma_{22} \end{vmatrix} = 0 \tag{3.25}$$

The elements of this coefficients matrix are written by:

$$\Gamma_{11} = -S_O + \sum_{n=1}^{\infty} \frac{2D_z L n^2\pi^2(L^2 + g^2n^2\pi^2)m\omega^2}{D_z n^4\pi^4(L^2 + g^2n^2\pi^2) - L^6 m\omega^2} \tag{3.26}$$

$$\Gamma_{12} = -\sum_{n=1}^{\infty} \frac{2(-1)^n D_z L n^2\pi^2(L^2 + g^2n^2\pi^2)m\omega^2}{D_z n^4\pi^4(L^2 + g^2n^2\pi^2) - L^6 m\omega^2} \tag{3.27}$$

$$\Gamma_{21} = -\sum_{n=1}^{\infty} \frac{2(-1)^n D_z L n^2\pi^2(L^2 + g^2n^2\pi^2)m\omega^2}{D_z n^4\pi^4(L^2 + g^2n^2\pi^2) - L^6 m\omega^2} \tag{3.28}$$

$$\Gamma_{22} = -S_L + \sum_{n=1}^{\infty} \frac{2D_z L n^2\pi^2(L^2 + g^2n^2\pi^2)m\omega^2}{D_z n^4\pi^4(L^2 + g^2n^2\pi^2) - L^6 m\omega^2} \tag{3.29}$$

As will be understood, the elements of the coefficients matrix used to obtain the frequencies include the material characteristic length and the stiffnesses of the elastic springs. These elastic spring stiffnesses can be set small enough to be zero or large enough to be full rigid. If the stiffnesses of these springs are set sufficiently small, free end conditions are satisfied, while if they are set sufficiently large, simply-supported end conditions are satisfied. In short, this solution has the ability to calculate the frequencies of nanobeams with both rigid and deformable boundaries.

3.4 Numerical examples

In this part of the work, we will present a comparison study showing the accuracy of the solution and graphs analyzing the frequencies of FG nanobeams. The component properties of the FG nanobeam are as follows [30]: $E_1 = 393$ GPa, $E_2 = 200$ GPa, $\rho_1 = 3400$ kg m^{-3}, $\rho_2 = 7800$ kg m^{-3}. A comparison study for homogeneous nanobeams is presented in table 3.1. The comparison study is performed with dimensionless frequencies (ϖ) and the following parameters are used for a consistent comparison: $p = 0$, $g = L/10$, $\overline{S_0} = \overline{S_L} = 10^{11}$. Here $\overline{S_0}$ and $\overline{S_L}$ are dimensionless spring parameters and the dimensionless expressions for the comparison study are assumed as follows:

$$\varpi = \omega L^2 \sqrt{\frac{\rho_1 A}{E_1 I}} \tag{3.30}$$

Table 3.1. Comparison study for simply-supported strain gradient beam.

	Strain gradient theory		Classical theory	
Mode number	[22]	This study	[22]	This study
1	10.3452	10.3452	9.8696	9.8696
2	46.6244	46.6244	39.45	39.4784
3	122.0600	122.0600	88.8264	88.8264
4	253.6043	253.604	157.914	157.914
5	459.4597	459.454	246.74	246.74
6	758.3004	758.148	335.306	335.306

$$\overline{S_0} = \frac{S_0 L^3}{E_1 I} \tag{3.31}$$

$$\overline{S_L} = \frac{S_L L^3}{E_1 I} \tag{3.32}$$

Since the comparison study shown in table 3.1 is given for a simply-supported nanobeam, the spring stiffnesses are assigned too high. Also, $n = 10$ is considered while solving the problem. Even with a low number of terms, the agreement is excellent. We will now proceed to the analysis for the functionally graded nanobeam. Unless otherwise stated, the parameters used are as follows: $\overline{S_0} = \overline{S_L} = 10^{11}$, $b = h = 1$ nm, $L = 20$ nm, $n = 10$, $p = 1$, $g = 4$. Also the dimensionless spring parameters used in the analysis are calculated as follows:

$$\overline{S_0} = \frac{S_0 L^3}{E_2 I} \tag{3.33}$$

$$\overline{S_L} = \frac{S_L L^3}{E_2 I} \tag{3.34}$$

Figure 3.2 shows the effect of material characteristic length. In this figure drawn for the first two modes, $g = 0,\ 1,\ 2,\ 3,\ 4$ nm are considered. It is clear that as the value of the material characteristic length increases, the frequencies also increase. This increase is more noticeable in the second mode and at higher g values. It should also be noted that at $g = 0$ the results belong to the classical theory.

Figure 3.3 shows the influence of the power-law index. In this figure drawn for classical theory and strain gradient theory, $p = 0,\ 0.5,\ 1,\ 1.5,\ 2$ are considered. It is clear that as the value of p increases, a significant decrease in frequencies occurs. This decrease is valid for both theories and occurs at the same rates. Also, it should be noted that at $p = 0$ the results give the frequencies of a homogeneous nanobeam composed of the top material.

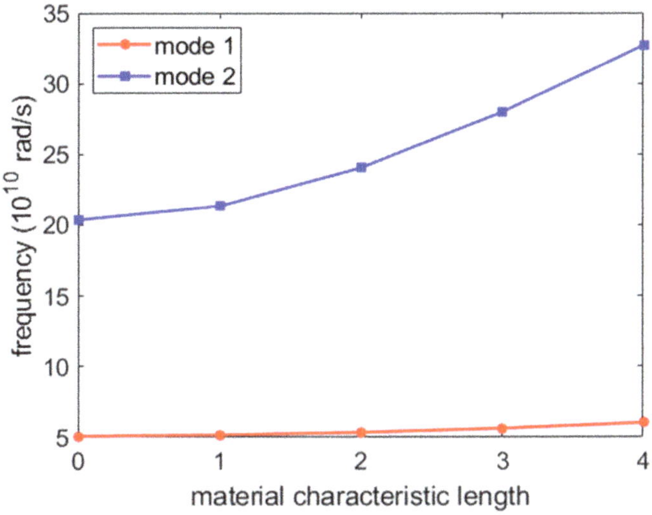

Figure 3.2. Effect of material characteristic length.

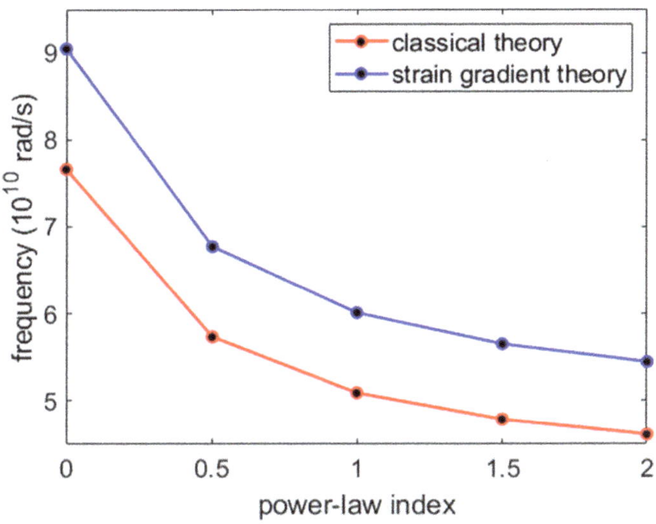

Figure 3.3. Effect of power-law index.

Figure 3.4 shows the effect of $\overline{S_0} = \overline{S_L}$. As mentioned, the uniqueness of this work is that it gives the results of deformable boundaries in the same solution. And again, as mentioned, we can adjust the spring stiffnesses to any value. In this figure drawn for classical theory and strain gradient theory, $\overline{S_0} = \overline{S_L} = 250,\ 500,\ 750,\ 1000,\ 1250$ are considered and the solutions are realized with 80 terms ($n = 80$). It is seen that as the value of $\overline{S_0} = \overline{S_L}$ increases, that is, as the stiffness of the springs increases, the frequencies increase. This increase is valid for both theories and it can be seen that

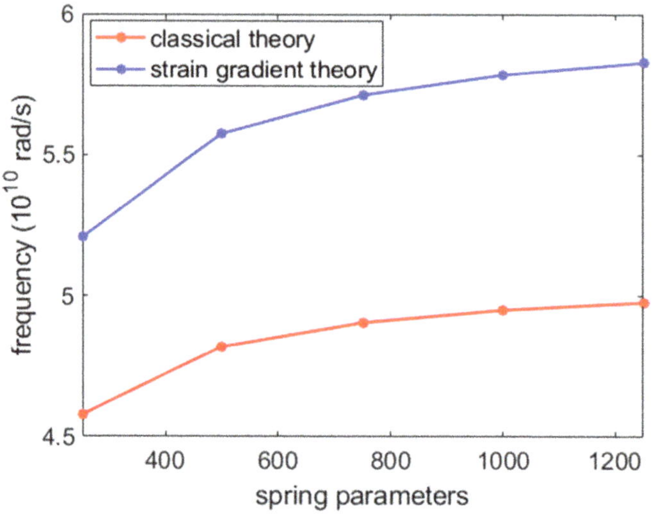

Figure 3.4. Effect of spring parameter.

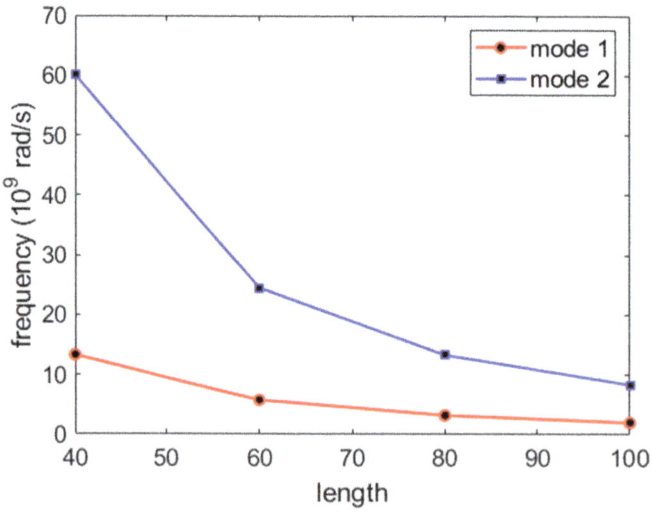

Figure 3.5. Effect of length.

the rate of increase is higher in the SGET. In other words, spring stiffnesses are more important in the analysis with SGET.

Figure 3.5 shows the effect of nanobeam length. In this figure drawn for the strain gradient theory, $L = 40,\ 60,\ 80,\ 100$ nm is considered. It can be seen that as the nanobeam length increases, i.e. the slenderness ratio increases, there is a significant decrease in frequencies. This decrease is valid for both modes. It is also noticeable that the frequencies of mode 1 and mode 2 converge as the length increases.

3.5 Conclusions

The aim of this study is to present an eigenvalue solution that finds the vibration frequencies of functionally graded nanobeams modeled with elastic lateral springs based on strain gradient theory. In the solution performed by Fourier sine series and Stokes' transform, the frequencies of both the simply-supported ends, which is a rigid boundary condition, and the deformable boundary states, can be calculated. First, the accuracy of the solution is demonstrated with a study found in the literature. Then, the effects of various parameters such as material characteristic length, power-law index, spring parameters, and length of the nanobeam on the frequencies of functionally graded nanobeams are examined with a number of figures. In addition, examples comparing the frequencies of both classical theory and strain gradient theory are also included.

References

[1] Zhang C, Chen F, Huang Z, Jia M, Chen G, Ye Y and Lavernia E J 2019 Additive manufacturing of functionally graded materials: a review *Mater. Sci. Eng.* A **764** 138209

[2] Toudehdehghan A, Lim J W, Foo K E, Ma'Arof M I N and Mathews J 2017 A brief review of functionally graded materials *MATEC Web Conf.* **vol 131** (Les Ulis: EDP Sciences)

[3] Sina S A, Navazi H M and Haddadpour H 2009 An analytical method for free vibration analysis of functionally graded beams *Mater. Des.* **30** 741–7

[4] Li X F, Kang Y A and Wu J X 2013 Exact frequency equations of free vibration of exponentially functionally graded beams *Appl. Acoust.* **74** 413–20

[5] Cao D, Gao Y, Yao M and Zhang W 2018 Free vibration of axially functionally graded beams using the asymptotic development method *Eng. Struct.* **173** 442–8

[6] Karamanlı A 2018 Free vibration analysis of two directional functionally graded beams using a third order shear deformation theory *Compos. Struct.* **189** 127–36

[7] Aldousari S M 2017 Bending analysis of different material distributions of functionally graded beam *Appl. Phys.* A **123** 296

[8] Li J, Guan Y, Wang G, Zhao G, Lin J, Naceur H and Coutellier D 2018 Meshless modeling of bending behavior of bi-directional functionally graded beam structures *Composites* B **155** 104–11

[9] Thai H T and Vo T P 2012 Bending and free vibration of functionally graded beams using various higher-order shear deformation beam theories *Int. J. Mech. Sci.* **62** 57–66

[10] Chaabane L A 2019 Analytical study of bending and free vibration responses of functionally graded beams resting on elastic foundation *Struct. Eng. Mech. Int. J.* **71** 185–96

[11] Huang Y 2020 Bending and free vibrational analysis of bi-directional functionally graded beams with circular cross-section *Appl. Math. Mech.* **41** 1497–516

[12] Hadji L and Bernard F 2020 Bending and free vibration analysis of functionally graded beams on elastic foundations with analytical validation *Adv. Mater. Res.* **9** 63

[13] Fleck N A and Hutchinson J 2001 A reformulation of strain gradient plasticity *J. Mech. Phys. Solids* **49** 2245–71

[14] Hutchinson J and Fleck N 1997 Strain gradient plasticity *Adv. Appl. Mech.* **33** 295–361

[15] Koiter W 1964 Couple-stresses in the theory of elasticity, I and II *Proc. Roy. Netherlands Acad. Sci.* B **67** 0964

[16] Lam D C, Yang F, Chong A C M, Wang J and Tong P 2003 Experiments and theory in strain gradient elasticity *J. Mech. Phys. Solids* **51** 1477–508
[17] Lim C W, Zhang G and Reddy J 2015 A higher-order nonlocal elasticity and strain gradient theory and its applications in wave propagation *J. Mech. Phys. Solids* **78** 298–313
[18] Yang F A C M, Chong A C M, Lam D C C and Tong P 2002 Couple stress based strain gradient theory for elasticity *Int. J. Solids Struct.* **39** 2731–43
[19] Eringen A C 1983 On differential equations of nonlocal elasticity and solutions of screw dislocation and surface waves *J. Appl. Phys.* **54** 4703–10
[20] Eringen A C and Edelen D 1972 On nonlocal elasticity *Int. J. Eng. Sci.* **10** 233–48
[21] Papargyri-Beskou S, Tsepoura K G, Polyzos D and Beskos D 2003 Bending and stability analysis of gradient elastic beams *Int. J. Solids Struct.* **40** 385–400
[22] Artan R and Batra R C 2012 Free vibrations of a strain gradient beam by the method of initial values *Acta Mech.* **223** 2393–409
[23] Artan R and Toksöz A 2013 Stability analysis of gradient elastic beams by the method of initial value *Arch. Appl. Mech.* **83** 1129–44
[24] Ramezani S 2012 A micro scale geometrically non-linear Timoshenko beam model based on strain gradient elasticity theory *Int. J. Non Linear Mech.* **47** 863–73
[25] Yayli M O 2011 Stability analysis of a gradient elastic beam using finite element method *Int. J. Phys. Sci.* **6** 2844–51
[26] Gül U 2023 Kısa Elyaf Hizalanmasının Kompozit Nanokirişlerin Burkulmasına Etkisi *23rd Ulusal Mekanik Kongresi (Konya Teknik Üniversitesi, 4–7 Eylül 2023)*
[27] Eltaher M A, Alshorbagy A E and Mahmoud F F 2013 Determination of neutral axis position and its effect on natural frequencies of functionally graded macro/nanobeams *Compos. Struct.* **99** 193–201
[28] Kim H K and Kim M S 2001 Vibration of beams with generally restrained boundary conditions using Fourier series *J. Sound Vib.* **245** 771–84
[29] Yaylı M Ö, Uzun B and Deliktaş B 2022 Buckling analysis of restrained nanobeams using strain gradient elasticity *Waves Random Complex Medium* **32** 2960–79
[30] Czechowski L 2015 Study of dynamic buckling of FG plate due to heat flux pulse *Int. J. Appl. Mech. Eng.* **20** 19–31

Chapter 4

Effect of the micromechanical models on the vibration of FGM nanobeams

Tlidji Youcef, Benferhat Rabia, Draiche Kada and Hassaine Daouadji Tahar

Symbols

E	Young's modulus
E_c	Young's modulus on the top
E_m	Young's modulus on the bottom
ρ	Mass density
ρ_c	Mass density on the top
ρ_m	Mass density on the bottom
ν	The Poisson's ratio
δ	Variational
L	Beam length
h	Beam thickness
Π	Strain energy
T	kinetic energy
k	The material parameters
$V(z)$	The volume fraction
ω	The frequency parameter
$\bar{\omega}$	The dimensionless frequency parameter
μ	The nonlocal parameter
U	Axial displacement components
W	Transverse displacement components
u_0	Axial displacement components of a point on the beam's neutral axis
w	Transverse displacement components of a point on the beam's neutral axis
$'$	Partial differentiation of the quantities to x.

doi:10.1088/978-0-7503-6024-1ch4

(*Continued*)

λ_i	Eigenvalues associated with the matrix T
$[E]$	Eigenvectors associated with the matrix T
T	Matrix
K	Constant column vector

4.1 Introduction

The predominant approach in structural analysis of functionally graded materials (FGMs) has been to employ Voigt and Mori–Tanaka micromechanical models. However, increasing attention is being paid to the advantages and disadvantages of both the rule of mixture and micromechanical models for structural analysis of FGMs. In the literature, various numerical and analytical methods are proposed to analyze the vibration of functionally graded (FG) nanobeams. Eltaher *et al* [1] presented vibration analysis of an Euler–Bernoulli FGM nanobeam using a Voigt micromechanical models and the finite element method. Farzad *et al* [2] employed the nonclassical beam model within the framework of the Euler–Bernoulli beam theory (EBT) to analyse the vibration of FG nanobeams, the differential transformation method is employed to solve the governing equations derived through Hamilton's principle. Javad *et al* [3] explores the influence of classical and non-classical boundary conditions on the free vibration characteristics of FG size-dependent nanobeams. A semi-analytical approach based on the differential transform method (DTM) is introduced for the first time in this context. Three mathematical models—power-law (P-FGM), sigmoid (S-FGM), and Mori–Tanaka (MT-FGM) distribution—are employed to depict the material properties across the thickness direction. Jena and Chakraverty [4] use the DTM to investigate the free vibration of nanobeams based on nonlocal EBT. Zhang *et al* [5] conducted a dynamic analysis of a FGM nanobeam, integrating the nonlocal theory with the first-order shear deformation theory. They utilized the precise-constant method in conjunction with the extended Wittrick–Williams algorithm to determine the structural frequencies of the nanobeam. Jena *et al* [6] explored the vibration characteristics of a FG porous nanobeam embedded within an elastic Winkler–Pasternak foundation. An EBT, is employed to describe the displacement of the FG nanobeam. To account for the small-scale effects of the FG nanobeam, a Bi-Helmholtz type of nonlocal elasticity is utilized. Additionally, the nanobeam is assumed to possess uniform porosity distributed evenly throughout its cross-section. The variation in Young's modulus and mass density of the nanobeam along its thickness, is modeled using a power-law model. Berghouti *et al* [7] conducted a study on the free vibration analysis of FG porous nanobeams, employing a nonlocal nth-order shear deformation theory. They use a power-law scheme to describe the variation of the material characteristics of the FG beam. Ghali *et al* [8] investigated the damping characteristics of free vibrations in FG viscoelastic nanobeams supported by viscoelastic foundations. They employed the infinite Kelvin–Voigt model within the framework of the Winkler–Pasternak type, incorporating the nonlocal strain-driven gradient elasticity theory proposed by Eringen. To account for the continuous variation in material properties in FGM nanobeams, a power-law model was adopted. The study utilized the first-order shear deformation theory, incorporating shear correction factors, and employed the finite element method for analysis.

This study shifts its focus toward investigating the free vibration response of an inhomogeneous nano-sized beam, incorporating various micromechanical models including Reuss, Tamura, and LRVE. The analysis of the nanobeam is conducted using classical beam theory, which is coupled with the general strain gradient theory through Hamilton's principle. The analytical solution is then obtained employing the state space approach and the differential quadrature method (DQM). Numerical simulations are subsequently carried out to demonstrate the influence of size-dependent effects on the vibration behavior of the nanobeam under different micromechanical models.

4.2 Theoretical formulation

Considering a straight FG nanobeam made from a mixture of aluminum (Al) and alumina (Al_2O_3). The nanobeam, having length (L) and thickness (h) figure 4.1.

4.2.1 Effective properties of FGMs

Voigt model

The Voigt model, that is frequently used in most FGM analyses, estimates Young's modulus (E) of FGMs as:

$$E(z) = (E_c - E_m)V(z) + E_m \tag{4.1}$$

where the subscripts c and m represent ceramic and metal. And the volume fraction $V(z)$ can be expressed, for power-law FG beam as [4, 9]:

$$V(z) = \left(\frac{1}{2} + \frac{z}{h}\right)^k \tag{4.2}$$

Reuss model

The Reuss model suggests that when homogeneous phases experience equivalent uniform stress at a macroscopic scale, their average stress is predicted to be identical. Consequently, this model details the expected Young's modulus (E) for FGMs as follows [10, 11]:

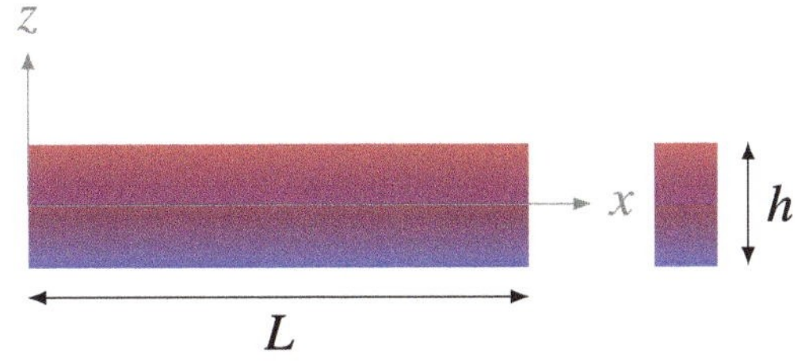

Figure 4.1. Geometry and coordinates of functionally graded beam.

$$E(z) = \frac{E_c E_m}{E_c(1 - V(z)) + E_m V(z)} \tag{4.3}$$

Tamura model

Tamura method, adopts a linear rule of mixture for determining the effective Young's modulus of a two-phase composite. This approach incorporates an empirical fitting parameter, q_T (stress-to-strain transfer), into the formulation of the effective Young's modulus. This empirical parameter establishes a relationship between stress and strain in both the matrix and particle phases. As a result, the resulting effective Young's modulus and Poisson's ratio are determined as [10, 11]:

$$E(z) = \frac{(1 - V(z))E_m(q_T - E_c) + V(z)E_c(q_T - E_m)}{(1 - V(z))(q_T - E_c) + V(z)(q_T - E_m)} \tag{4.4}$$

Voigt and Reuss schemes can be obtained within the Tamura formulation by setting specific values for qT. Specifically, the Voigt scheme is achieved by setting $q_T = +\infty$, while the Reuss scheme is obtained by setting $q_T = 0$.

Cubic local representative volume elements (LRVE) model

In the cubic LRVE to derive the effective material properties of FGMs, we assume the second phase of a two-phase composite as an inclusion. In this micromechanical model, the Young's modulus is expressed as [10, 11]:

$$E(z) = E_m(1 - \sqrt[3]{V(z)})\left(1 - \frac{1}{1 - \sqrt[3]{V(z)}\left(1 - \frac{E_m}{E_c}\right)}\right) \tag{4.5}$$

4.3 Kinematics

Assuming that the deformation of FG beam is only in (x, z) plane and let $U(x, z)$ and $W(x, z)$ be the axial and transverse displacement components of an arbitrary point. These components can be expressed as shown in equation (4.6):

$$U(x, z) = u_0(x) - zw'(x) \tag{4.6a}$$

$$W(x, z) = w(x) \tag{4.6b}$$

where u_0 and w present the displacement components of a point on the beam's neutral axis along x and z directions. For brevity, prime (′) represents the partial differentiation of the quantities to x.

The strains related to the displacement field in equation (4.6) are:

$$\epsilon_x = \frac{\partial U(x, z)}{\partial x} = u_0'(x) - zw''(x) \tag{4.7}$$

4.4 Governing equations of motions

In order to derive the governing equations of motions and boundary conditions, Hamilton's principle is employed.

$$\int \delta(\Pi - T)\mathrm{d}v = 0 \tag{4.8}$$

where $\delta\Pi$ and δT denote the virtual variation of the strain energy and kinetic energy. The virtual variation of the strain energy is given by:

$$\delta\Pi = \int_{-\frac{L}{2}}^{\frac{L}{2}} \int_{0}^{b} \int_{-\frac{h}{2}}^{\frac{h}{2}} [\sigma_x \delta\epsilon_x]\mathrm{d}z\mathrm{d}y\mathrm{d}x \tag{4.9a}$$

$$= \int_{-\frac{L}{2}}^{\frac{L}{2}} [N_x \delta u_0 - M_x \delta w'']\mathrm{d}x \tag{4.9b}$$

in which:

$$N_x = \int_{-\frac{h}{2}}^{\frac{h}{2}} \sigma_x b\mathrm{d}z \tag{4.10a}$$

$$M_x = \int_{-\frac{h}{2}}^{\frac{h}{2}} \sigma_x zb\mathrm{d}z \tag{4.10b}$$

The kinetic energy for an Euler–Bernoulli beam can be expressed as:

$$\delta T = \int_{-\frac{L}{2}}^{\frac{L}{2}} \rho(z)\left[\left(\frac{\partial U(x, t)}{dt}\right)^2 + \left(\frac{\partial W(x, t)}{dt}\right)^2\right]\mathrm{d}x \tag{4.11}$$

the virtual kinetic energy can be expressed as

$$\begin{gathered}\delta T = \int_{-\frac{L}{2}}^{\frac{L}{2}} I_0\left(\frac{\partial u(x, t)}{\partial t}\frac{\partial \delta u(x, t)}{\partial t}\right) + \left(\frac{\partial w(x, t)}{\partial t}\frac{\partial \delta w(x, t)}{\partial t}\right) - \\ I_1\left(\frac{\partial u(x, t)}{\partial t}\frac{\partial^2 \delta w(x, t)}{\partial t \partial x}\right) + \left(\frac{\partial^2 w(x, t)}{\partial t \partial x}\frac{\partial \delta u(x, t)}{\partial t}\right) + I_2\left(\frac{\partial^2 w(x, t)}{\partial t \partial x}\frac{\partial^2 \delta w(x, t)}{\partial t \partial x}\right)\mathrm{d}x\end{gathered} \tag{4.12}$$

where the mass moments of inertia are:

$$(I_0, I_1, I_2) = \int_{-\frac{h}{2}}^{\frac{h}{2}} \rho(z)(1, z, z^2)\mathrm{d}z \tag{4.13}$$

where

$$\rho(z) = (\rho_c - \rho_m)V(z) + \rho_m \tag{4.14}$$

By substituting equations (4.9) and (4.12) into equation (4.8) and setting the coefficients of δu and δw to zero, the following Euler–Lagrange equation can be obtained

$$\delta u_0\colon N_x' = I_0\ddot{u} - I_1\ddot{w}' \tag{4.15a}$$

$$\delta w\colon M''_x = I_0\ddot{w} + I_1\ddot{u}' - I_2\ddot{w}'' \tag{4.15b}$$

For simplification, the mass moment of inertia terms in the equation (4.15) are assumed to be negligible, which results in a simplified version of Euler–Lagrange equation as:

$$\delta u_0\colon N_x' = 0 \tag{4.16a}$$

$$\delta w\colon M''_x = I_0\ddot{w} + I_1\ddot{u}' - I_2\ddot{w}'' \tag{4.16b}$$

4.4.1 The nonlocal elasticity model for FG nanobeam

According to Eringen the stress at a reference point is a functional of the strain field at every point of the continuum. Eringen proposed a differential form of the nonlocal constitutive relation as [4]:

$$\sigma_{xx} - \mu\frac{\partial^2\sigma_{xx}}{\partial x^2} = E(z)\epsilon_{xx} \tag{4.17}$$

where ($\mu = (e_0 a)^2$), The parameter $e_0 a$ is the scale coefficient revealing the small scale effect on the responses of structures of nano size. Integrating equation (4.17) over the beam's cross-section area, we obtain the force-strain and the moment-strain of the nonlocal Euler–Bernoulli FG beam theory can be obtained as follows

$$N - \mu N'' = Au' - Bw'' \tag{4.18a}$$

$$M - \mu M'' = Bu' - Cw'' \tag{4.18b}$$

where

$$(A, B, C) = \int_{-\frac{h}{2}}^{\frac{h}{2}} E(z)(1, z, z^2)\mathrm{d}z \tag{4.19}$$

By using equation (4.15) it is concluded that N is a constant value throughout the beam; therefore, it can be expressed as

$$N = Au' - Bw'' \tag{4.20}$$

Also the explicit relation of the nonlocal bending moment can be derived by substituting for the second derivative of M from equation (4.15) into equation (4.18) as follows

$$M = Bu' - Cw'' + \mu(I_0\ddot{w} + I_1\ddot{u}' - I_2\ddot{w}'') \tag{4.21}$$

The nonlocal governing equations of Euler–Bernoulli FG nanobeam in terms of the displacement can be derived as follows.

$$\frac{B^2}{A}w^{(iv)} - Cw^{(iv)} + \mu\left(I_0\ddot{w}'' + \frac{B\ I_1}{A}\ddot{w}^{(iv)} - I_2\ddot{w}\right) - I_0\ddot{w} - \frac{B\ I_1}{A}\ddot{w}'' + I_2\ddot{w}'' = 0 \quad (4.22)$$

4.5 Solution of the governing equation

Displacement components can be expressed as:

$$w(x,\ t) = W(x)e^{i\omega} \quad (4.23)$$

where ω is the natural frequency of a FGM nanobeam.

Substituting equation (4.23) into equation (4.22) leads to:

$$\begin{aligned} &\frac{B^2}{A}w^{(iv)} - Cw^{(iv)} + \mu\left(-\omega^2\ I_0 w'' - \frac{\omega^2\ B\ I_1}{A}w^{(iv)} + \omega^2\ I_2 w\right) \\ &+ \omega^2\ I_0 w + \frac{\omega^2\ B\ I_1}{A}w'' - \omega^2\ I_2 w'' = 0 \end{aligned} \quad (4.24)$$

4.5.1 State space approach

Using the state space approach [9, 12], an ordinary differential equation is obtained, form the equation (4.24) as:

$$W^{iv} = b_1\ W' + b_2\ W'' \quad (4.25)$$

where the b_n are the constants coefficients: $b_1 = \frac{-CA_s}{AH - C^2}$, $b_2 = \frac{BH - CF}{AH - C^2}$.

The systems of equations (4.25) can be converted into a matrix form as:

$$Z'(x) = TZ(x) \quad (4.26)$$

where

$$Z(x) = \{W,\ W',\ W',\ W'''\} \quad (4.27)$$

And matrix $[T]$ is defined as

$$[T] = \begin{bmatrix} 0 & 1 & 0 & 0 \\ 0 & 0 & 1 & 0 \\ 0 & 0 & 0 & 1 \\ b_1 & 0 & b_2 & 0 \end{bmatrix} \quad (4.28)$$

A formal solution of equation (4.26) is given by:

$$Z(x) = e^{Tx}\ K \quad (4.29)$$

K is a constant column vector determined from the boundary conditions at $x = \pm\frac{L}{2}$ and e^{Tx} is the general matrix solution

$$e^{Tx} = [E]\begin{bmatrix} e^{\lambda_1 x} & & 0 \\ & \vdots & \\ 0 & & e^{\lambda_i x} \end{bmatrix}[E]^{-1} \tag{4.30}$$

λ_i, $i = 1, \ldots, 4$ are eigenvalues and $[E]$ eigenvectors, associated with the matrix $[T]$.

Boundary conditions

The unknown function $Z(x)$ can be used to express the boundary conditions as follows

$$\text{Clamped}: W = W' = 0 \tag{4.31a}$$

$$\text{Hinged}: W' = W'' = 0 \tag{4.31b}$$

$$\text{Free}: W'' = W''' = 0 \tag{4.31c}$$

Substituting equation (4.26) into equation (4.31), a homogeneous system of equations is obtained as:

$$e^{Tx}K = 0 \tag{4.32}$$

The nontrivial solutions for equation (4.32) require the determinant of $|e^{Tx}|$ being zero, from which the natural frequency can be determined. It should be noted that a trial and error procedure need to be used to obtain the natural frequency values due to the attendant of unknown ω in matrix $[T]$.

4.5.2 Differential quadrature formulation

Polynomials expressed as weighted linear sums of function values at a grid of preselected discrete points are used in differential quadrature to approximate the subspace derivatives of functions in differential equations [13, 14]. The differential quadrature approximation of the ith sampling point's mth order derivative is given by for a function $f(x)$

$$\frac{d^m}{dx^m}\begin{Bmatrix} f(x_1) \\ f(x_2) \\ \cdot \\ f(x_n) \end{Bmatrix} = c_{ij}^{(m)}\begin{Bmatrix} f(x_1) \\ f(x_2) \\ \cdot \\ f(x_n) \end{Bmatrix} \quad \text{for} \quad i, j = 1, 2, \ldots, n \tag{4.33}$$

where the number of sampling points is the value n assuming a polynomial for Lagrangian interpolation

$$f(x) = \frac{M(x)}{(x - x_i)M_1(x_i)} \quad \text{for } i, j = 1, 2, \ldots, n \tag{4.34}$$

where

$M(x) = \prod_{j=1}^{n} (x - x_j)$

$M_1(x_i) = \prod_{j=1, j\neq i}^{m} (x_i - x_j)$ for $i, j = 1, 2, \ldots, n$

Substituting in equation (4.34) leads to

$$C_{ij}^{(1)} = \frac{M_1(x_1)}{(x_i - x_i)M_1(x_j)} \text{ for } i, j = 1, 2, \ldots, n \tag{4.35}$$

$$C_{ii}^{(1)} = -\sum_{j=1, j\neq i}^{n} C_{IJ}^{(1)} \text{ for } i, j = 1, 2, \ldots, n \tag{4.36}$$

It is possible to determine the higher derivative as

$$C_{ii}^{(m)} = -\sum_{j=1, j\neq i}^{n} C_{ik}^{(1)} C_{kj}^{(m-1)} \text{ for } i, j = 1, 2, \ldots, n \tag{4.37}$$

For the sampling points, we adopt well accepted Chebyshev–Gauss–Lobatto mesh distribution

$$x_i = \frac{1}{2}\left[1 - \cos\frac{(i-1)}{(n-1)}\pi\right] \tag{4.38}$$

where n is the number of sampling points.

Assuming $c(:, :, m)$ is the mth derivative, equation (4.24) may be written as

$$\left[\left(\frac{B^2}{A} - C - \frac{\omega^2\ B\ I_1}{A}\right)c(:, :, 4) - \left(\mu\omega^2\ I_0 + \frac{\omega^2\ B\ I_1}{A} - \omega^2\ I_2\right)c(:, :, 2) + (\mu\omega^2\ I_2 + \omega^2\ I_0)c(:, :, 1)\right]\{w\} = 0 \tag{4.39}$$

4.6 Numerical results and discussion

Tables and figures showing the dimensionless frequency parameters, $\bar{\omega} = \omega\frac{L^2}{h}\sqrt{\frac{\rho_m}{E_m}}$, of a FGM nanobeam made from aluminum (Al) and alumina (Al_2O_3). The nanobeam has length (L) = 10 000 nm, width (b) = 1000 nm and thickness (h) = 100 nm, with different boundary conditions simply-supported/simply-supported, (SS), clamped-clamped (CC), clamped-simply-supported (CS), and clamped-free (CF). The following material properties are employed here: $E_m = 70$ GPa, $\rho_m = 2702$ kg m^{-3} and $E_c = 380$ GPa, $\rho_c = 3960$ kg m^{-3} and $\nu = 0.3$.

4.6.1 Convergence of DQM

A convergence study is carried out for the DQ method. Tables 4.1 and 4.2 present the convergence analysis of the DQM for the first three natural frequencies of a nanobeam, considering various gradient indexes and classical boundary conditions. The results indicate that after a certain number of iterations, the eigenvalues converge to precise values. Notably, tables 4.1 and 4.2 reveal a high convergence rate of the method. Specifically, it is observed that for a FG nanobeam with a gradient index of $p = 0.5$ and simply-supported boundary conditions the third

Table 4.1. Convergence of the first three natural frequencies of the P-FG nanobeam for ($L/h = 5$, $p = 0.5$, $\mu = 1$ nm).

BC	SS			CC		
N	Mode 1	Mode 2	Mode 3	Mode 1	Mode 2	Mode 3
4	4.5267	—	—	9.9495	—	—
6	4.3830	15.9677	29.7716	9.7430	24.1425	38.6747
8	4.3853	14.8603	26.8540	9.7463	22.1572	34.4253
10	4.3852	14.8971	26.9778	9.7463	22.1470	34.3272
12	4.3852	14.8957	26.9674	9.7463	22.1460	34.3213
14	4.3852	14.8957	26.9678	9.7463	22.1460	34.3215
16	4.3852	14.8957	26.9678	9.7463	22.1460	34.3215
18	4.3852	14.8957	26.9678	9.7463	22.1460	34.3215
SSA	4.3853	14.8957	26.9678	9.7463	22.1461	34.3215
Javad *et al* [3]	4.3852	14.8957	26.9678	9.7462	22.1460	34.3215

Table 4.2. Convergence of the first three natural frequencies of the P-FG nanobeam for ($L/h = 5$, $p = 0.5$, $\mu = 1$ nm).

BC	CF			CS		
N	Mode 1	Mode 2	Mode 3	Mode 1	Mode 2	Mode 3
4	1.4918	—	—	6.9587	—	—
6	1.6755	9.7180	24.8386	6.7770	19.8194	34.4173
8	1.6725	9.5251	22.5788	6.7715	18.3771	30.7311
10	1.6726	9.5334	22.1625	6.7716	18.4170	30.6362
12	1.6726	9.5333	22.1922	6.7716	18.4148	30.6385
14	1.6726	9.5333	22.1906	6.7716	18.4148	30.6379
16	1.6726	9.5333	22.1907	6.7716	18.4148	30.6379
18	1.6726	9.5333	22.1907	6.7716	18.4148	30.6379
SSA	1.6726	9.5333	22.1907	6.7716	18.4149	30.6379
Javad *et al* [3]	1.6725	9.5333	22.1906	6.7715	18.4149	30.6379

natural frequency converged after $n = 16$ number of grid points or terms with a precision of four digits. Similarly, the first and second natural frequencies achieved convergence after $n = 12$ and $n = 14$ terms, respectively. It seems that $n = 20$ is sufficient for all the considered modes and boundary conditions. Further, the results obtained from the DQM are validated with respect to the state space approach and the results of Javad *et al* [3].

Tables 4.3–4.5 depict the effect of the material graduation and the micro-mechanical models on first dimensionless frequency of a FG nanobeam with various boundary conditions, for slenderness ratios ($L/h = 5$). The nonlocal parameter is taken equal to $\mu = 0$, 2, and 4.

Table 4.3. Material graduation effect on first dimensionless frequency of a FG nanobeam with ($L/h = 5$, $\mu = 0$).

			Model				
Bcs	k	Method	Voigt	Reus	Trauma		LRVE
					$q = \infty$	$q = 0$	
CC	0.50	SSA	10.380 38	8.567 16	10.38039	8.567 16	9.210 99
		DQM	10.380 38	8.567 16	10.380 38	8.567 16	9.210 99
		Javad *et al* [3]	10.3804	—	—	—	—
	1	SSA	9.386 00	8.03236	9.385 99	8.032 36	8.445 04
		DQM	9.386 00	8.032 36	9.385 99	8.032 36	8.445 04
		Javad *et al* [3]	9.3859	—	—	—	—
	5	SSA	8.147 46	7.445 09	8.147 46	7.445 09	7.763 69
		DQM	8.14746	7.445 09	8.147 46	7.445 09	7.763 69
		Javad *et al* [3]	8.1474	—	—	—	—
SS	0.50	SSA	4.596 56	3.793 45	4.596 56	3.793 45	4.078 59
		DQM	4.596 55	3.79345	4.596 55	3.793 45	4.078 58
		Javad *et al* [3]	4.5965	—	—	—	—
	1	SSA	4.156 10	3.556 61	4.156 10	3.556 61	3.739 32
		DQM	4.156 09	3.556 60	4.156 09	3.556 60	3.739 31
		Javad *et al* [3]	4.1560	—	—	—	—
	5	SSA	3.608 34	3.297 50	3.608 34	3.297 50	3.438 50
		DQM	3.608 33	3.297 49	3.608 33	3.297 49	3.438 50
		Javad *et al* [3]	3.6083	—	—	—	—
CF	0.50	SSA	1.665 61	1.374 28	1.665 61	1.374 28	1.477 67
		DQM	1.665 61	1.374 28	1.665 61	1.374 28	1.477 67
		Javad *et al* [3]	1.6656	—	—	—	—
	1	SSA	1.505 78	1.288 41	1.505 78	1.288 42	1.354 57
		DQM	1.505 78	1.28841	1.505 78	1.288 41	1.354 57
		Javad *et al* [3]	1.6656	—	—	—	—
	5	SSA	1.308 41	1.196 08	1.308 41	1.196 07	1.247 05
		DQM	1.308 41	1.196 07	1.308 41	1.196 07	1.247 05
		Javad *et al* [3]	1.3084	—	—	—	—
CS	0.50	SSA	7.162 29	5.911 11	7.162 29	5.911 11	6.355 36
		DQM	7.162 29	5.911 10	7.162 29	5.911 10	6.355 35
		Javad *et al* [3]	7.1622	—	—	—	—
	1	SSA	6.476 11	5.542 09	6.47611	5.542 09	5.826 81
		DQM	6.47611	5.542 08	6.476 11	5.542 08	5.826 81
		Javad *et al* [3]	6.4761	—	—	—	—
	5	SSA	5.621 89	5.13736	5.621 89	5.137 36	5.357 15
		DQM	5.62188	5.137 35	5.621 88	5.137 35	5.357 14
		Javad *et al* [3]	5.6218	—	—	—	—

Table 4.4. Material graduation effect on first dimensionless frequency of a FG nanobeam with ($L/h = 5$, $\mu = 2$ nm).

Bcs	k	Method	Model				
			Voigt	Reus	Trauma		LRVE
					$q = \infty$	$q=0$	
CC	0.50	SSA	9.21394	7.605 34	9.213 94	7.605 34	8.176 64
		DQM	9.213 94	7.605 33	9.213 94	7.605 33	8.176 64
		Javad *et al* [3]	9.2139	—	—	—	—
	1	SSA	8.331 91	7.13075	8.331 91	7.130 76	7.497 21
		DQM	8.331 91	7.130 75	8.331 91	7.130 75	7.497 20
		Javad *et al* [3]	8.3319	—	—	—	—
	5	SSA	7.229 49	6.605 24	7.229 49	6.605 24	6.888 38
		DQM	7.22949	6.605 24	7.22949	6.605 24	6.888 37
		Javad *et al* [3]	7.2294	—	—	—	—
SS	0.50	SSA	4.200 63	3.466 70	4.200 63	3.466 70	3.727 27
		DQM	4.200 63	3.466 70	4.200 63	3.466 70	3.727 27
		Javad *et al* [3]	4.2006	—	—	—	—
	1	SSA	3.798 11	3.250 26	3.798 11	3.250 26	3.417 23
		DQM	3.798 11	3.250 25	3.798 11	3.250 25	3.417 22
		Javad *et al* [3]	3.7981	—	—	—	—
	5	SSA	3.297 53	3.013 47	3.297 53	3.013 47	3.142 32
		DQM	3.297 53	3.013 46	3.29753	3.013 46	3.142 32
		Javad *et al* [3]	3.2975	—	—	—	—
CF	0.50	SSA	1.679 67	1.385 88	1.679 67	1.385 88	1.490 14
		DQM	1.679 66	1.385 88	1.679 66	1.385 88	1.490 14
		Javad *et al* [3]	1.6796	—	—	—	—
	1	SSA	1.518 49	1.299 29	1.518 49	1.299 29	1.366 01
		DQM	1.518 49	1.299 29	1.518 49	1.299 29	1.366 00
		Javad *et al* [3]	1.5184	—	—	—	—
	5	SSA	1.319 44	1.206 14	1.319 44	1.206 14	1.257 55
		DQM	1.319 43	1.20614	1.319 43	.20614	1.257 55
		Javad *et al* [3]	1.3194	—	—	—	—
CS	0.50	SSA	6.437 71	5.313 29	6.437 71	5.313 29	5.71256
		DQM	6.437 71	5.313 29	6.437 71	5.313 29	5.712 56
		Javad *et al* [3]	6.4377	—	—	—	—
	1	SSA	5.82109	4.981 64	5.821 09	4.981 64	5.237 59
		DQM	5.82108	4.981 63	5.821 08	4.981 63	5.237 58
		Javad *et al* [3]	5.8210	—	—	—	—
	5	SSA	5.052 61	4.61692	5.052 61	4.616 92	4.814 54
		DQM	5.05260	4.616 91	5.052 60	4.616 91	4.814 54
		Javad *et al* [3]	5.0526	—	—	—	—

Table 4.5. Material graduation effect on first dimensionless frequency of a FG nanobeam with ($L/h = 5$, $\mu = 4$ nm).

Bcs	k	Method	Model				
			Voigt	Reus	Trauma		LRVE
					$q = \infty$	$q = 0$	
CC	0.50	SSA	8.36463	6.904 82	8.36464	6.904 82	7.423 35
		DQM	8.364 63	6.904 81	8.364 63	6.904 81	7.423 35
		Javad *et al* [3]	8.3646	—	—	—	—
	1	SSA	7.564 27	6.474 04	7.564 27	6.474 04	6.806 80
		DQM	7.564 26	6.474 04	7.564 26	6.474 04	6.806 80
		Javad *et al* [3]	7.5642	—	—	—	—
	5	SSA	6.561 69	5.994 51	6.561 69	5.994 51	6.251 74
		DQM	6.56169	5.994 51	6.561 69	5.994 51	6.251 74
		Javad *et al* [3]	6.5616	—	—	—	—
SS	0.50	SSA	3.892 06	3.212 04	3.892 06	3.212 04	3.453 48
		DQM	3.892 05	3.212 04	3.892 05	3.21204	3.453 47
		Javad *et al* [3]	3.8920	—	—	—	—
	1	SSA	3.519 10	3.011 50	3.519 10	3.011 50	3.166 20
		DQM	3.519 10	3.011 49	3.519 10	3.011 49	3.166 20
		Javad *et al* [3]	3.5191	—	—	—	—
	5	SSA	3.05530	2.792 10	3.055 30	2.792 10	2.911 50
		DQM	3.055 29	2.792 10	3.055 29	2.792 10	2.911 49
		Javad *et al* [3]	3.0552	—	—	—	—
CF	0.50	SSA	1.694 44	1.398 08	1.694 44	1.398 08	1.503 25
		DQM	1.69444	1.398 08	1.694 44	1.398 08	1.503 25
		Javad *et al* [3]	1.6944	—	—	—	—
	1	SSA	1.531 85	1.310 73	1.531 85	1.310 73	1.378 03
		DQM	1.531 85	1.310 73	1.531 85	1.310 73	1.378 03
		Javad *et al* [3]	1.5318	—	—	—	—
	5	SSA	1.331 02	1.216 72	1.331 02	1.216 72	1.268 59
		DQM	1.331 02	1.216 72	1.331 02	1.216 72	1.268 58
		Javad *et al* [3]	1.3310	—	—	—	—
CS	0.50	SSA	5.894 13	4.864 76	5.894 13	4.864 76	5.230 30
		DQM	5.894 12	4.864 76	5.894 12	4.864 76	5.230 29
		Javad *et al* [3]	5.8941	—	—	—	—
	1	SSA	5.32965	4.561 13	5.329 65	4.561 13	4.795 49
		DQM	5.32965	4.561 12	5.329 65	4.561 12	4.795 48
		Javad *et al* [3]	5.3296	—	—	—	—
	5	SSA	4.625 65	4.226 64	4.625 65	4.226 64	4.407 63
		DQM	4.62565	4.226 64	4.625 65	4.226 64	4.407 62
		Javad *et al* [3]	4.6256	—	—	—	—

For all boundary conditions and micromechanical models, the dimensionless frequency of a FG nanobeam decreases as the material graduation index increases. These tables reveal that the Voigt micromechanical model gives significantly higher values of $\bar{\omega}$ compared to the other micromechanical models.

Figure 4.2 shows the effect of aspect ratio $(\frac{L}{h})$ on the first dimensionless frequency parameters of FGM nanobeam with $\mu = 3$, under clamped-clamped, clamped-free

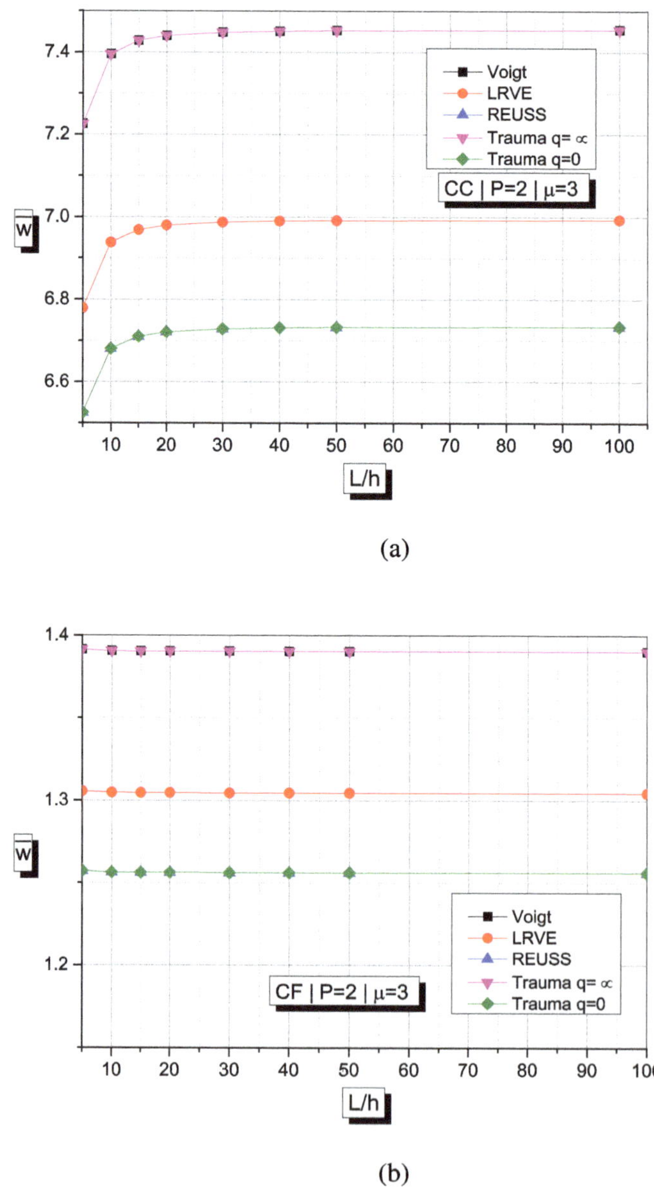

Figure 4.2. Variation of the dimensionless frequency parameter with aspect ratio $(\frac{L}{h})$.

boundary conditions and different $\frac{L}{h}$. From this figures, it can be seen that when $\frac{L}{h}$ increases, the dimensionless frequency parameter increases.

The analysis examines the effect of the nonlocal parameter on the first dimensionless vibrating mode ($\bar{\omega}$) of nanobeams. The nonlocal parameter values considered are 0, 1, 2, 3, and 4 nm, with boundary conditions including CC, and CF. the graphical results are presented in figures 4.3 and 4.4. It reveals a general decrease in

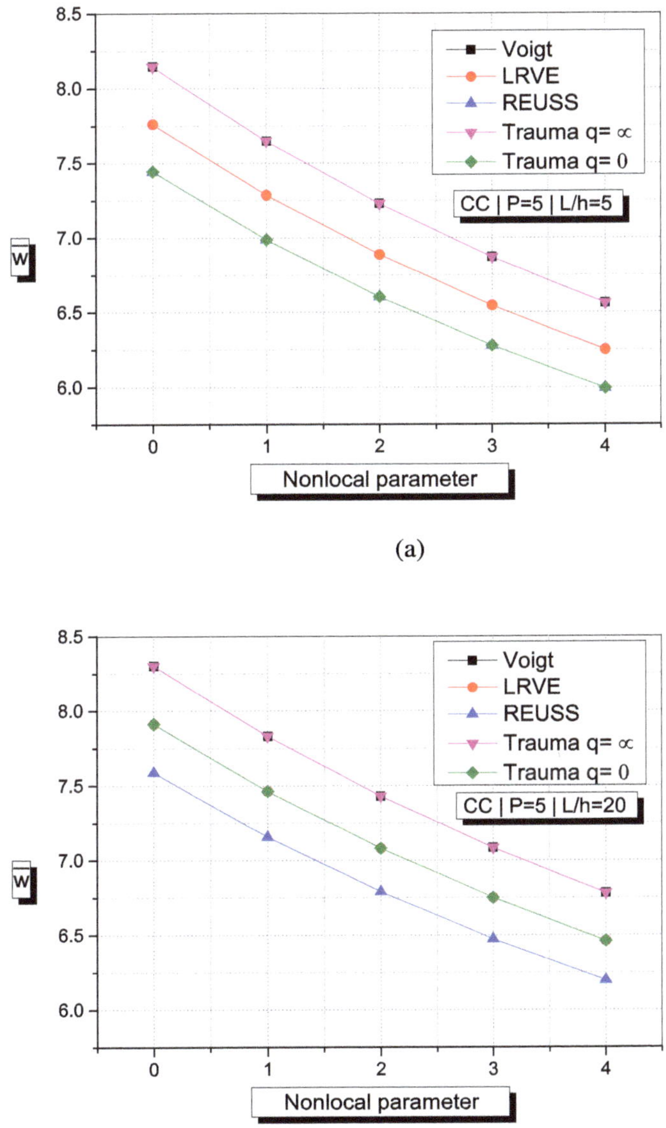

(a)

(b)

Figure 4.3. The variation of the dimensionless frequency parameter with nonlocal parameter for CC nanobeam.

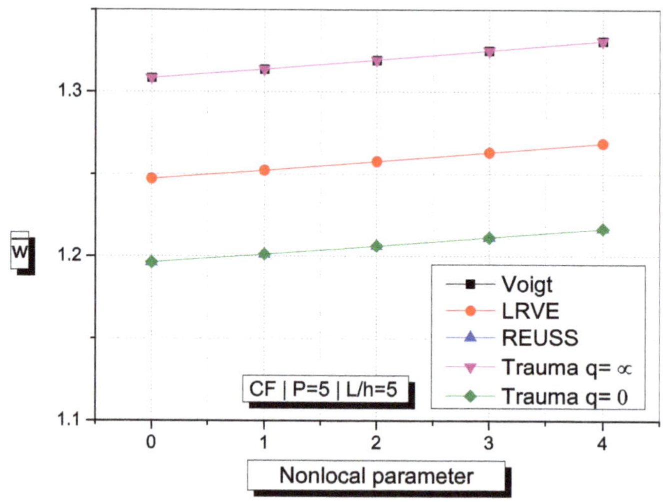

(a)

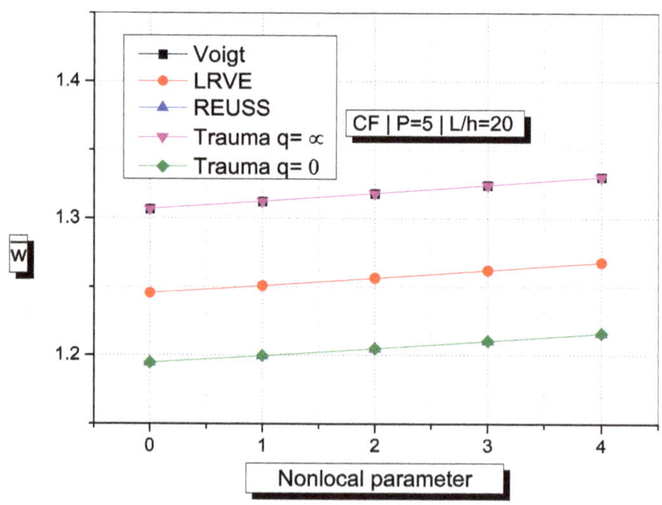

(b)

Figure 4.4. The variation of the dimensionless frequency parameter with nonlocal parameter for CF nanobeam.

frequency parameters with an increase in the nonlocal parameter, except for the first fundamental frequency parameter of CF nanobeams. CC nanobeams demonstrate notably higher frequency parameters at the edges compared to other boundary conditions.

4.7 Conclusion

This chapter explores the vibrational characteristics of FG nanobeams under different classical boundary conditions. The analysis is conducted using the framework of nonlocal elasticity theory combined with the application of DQM and the state space approach. The nanobeam is modeled using Eringen's theory of nonlocal elasticity in conjunction with EBT. Different micromechanical models are employed to assess the mechanical properties of FG nanobeams, where material properties change gradually across the thickness. The equations of motion are derived using Hamilton's principle and subsequently solved using DQM and the state space approach. In investigating the impact of micromechanical models on the free vibration analysis of FG nanobeams, our findings indicate a high level of agreement in frequency parameter predictions across the various micromechanical models. Any disparities observed could be attributed to differences in the calculation of Young's modulus within the models. It is observed from results that the the Voigt micromechanical model gives significantly higher values of $\bar{\omega}$ compared to the other micromechanical models.

References

[1] Eltaher M A, Emam S A and Mahmoud F F 2012 Free vibration analysis of functionally graded size-dependent nanobeams *Appl. Math. Comput.* **218** 7406–20

[2] Ebrahimi F, Ghadiri M, Salari E, Hoseini S A H and Shaghaghi G R 2015 Application of the differential transformation method for nonlocal vibration analysis of functionally graded nanobeams *J. Mech. Sci. Technol.* **29** 1207–15

[3] Ehyaei J, Ebrahimi F and Salari E 2016 Nonlocal vibration analysis of FG nano beams with different boundary conditions *Adv. Nano Res.* **4** 85–111

[4] Jena S K and Chakraverty S 2018 Free vibration analysis of Euler-Bernoulli nanobeam using differential transform method *Int. J. Comput. Mater. Sci. Eng.* **07** 1850020

[5] Zhang K, Ge M-H, Zhao C, Deng Z-C and Xu X-J 2019 Free vibration of nonlocal timoshenko beams made of functionally graded materials by symplectic method *Compos. Part B: Eng.* **156** 174–84

[6] Jena S K, Chakraverty S and Tornabene F 2019 Dynamical behavior of nanobeam embedded in constant, linear, parabolic, and sinusoidal types of Winkler elastic foundation using first-order nonlocal strain gradient model *Mater. Res. Express* **6** 0850f2

[7] Berghouti H, Adda Bedia E A, Benkhedda A and Tounsi A 2019 Vibration analysis of nonlocal porous nanobeams made of functionally graded material *Advances Nano Res.* **7** 351–64

[8] Drici G, Mechab I, Abbad H, Elmeiche N and Mechab B 2022 Investigating the free vibration of viscoelastic FGM timoshenko nanobeams resting on viscoelastic foundations with the shear correction factor using finite element method *Mech. Based Des. Struct. Mach.* **52** 1278–303

[9] Tlidji Y, Benferhat R, Trinh L C, Hassaine D T and Tounsi A 2021 New state state-space approach to dynamic analysis of porous FG beam under different boundary conditions *Adv. Nano Res.* **11** 347–59

[10] Akbarzadeh A H, Abedini A and Chen Z T 2015 Effect of micromechanical models on structural responses of functionally graded plates *Compos. Struct.* **119** 598–609

[11] Shahsavari D and Karami B 2022 Assessment of Reuss, Tamura, and LRVE models for vibration analysis of functionally graded nanoplates *Arch. Civ. Mech. Eng.* **22** 1–13

[12] Tlidji Y, Benferhat R, Draiche K and Hassaine D T 2023 Assessing the effects of porosity on the buckling of functionally graded beams *Functionally Graded Structures* (Bristol: IOP Publishing) 2053–563 pp pages 6-1 to 6-16

[13] Shu C 2012 *Differential Quadrature and its Application in Engineering* (Berlin: Springer Science)

[14] Jena S K, Chakraverty S, Mahesh V and Harursampath D 2022 Application of Haar wavelet discretization and differential quadrature methods for free vibration of functionally graded micro-beam with porosity using modified couple stress theory *Eng. Anal. Bound. Elem.* **140** 167–85

Chapter 5

Chebyshev polynomials based Rayleigh–Ritz method for free vibration analysis of axially functionally graded cantilever nanobeam

Akash Kumar Gartia and S Chakraverty

This study explores the free vibration analysis of an axially functionally graded (FG) cantilever nanobeam made of steel and alumina (Al_2O_3). A power-law distribution is used to represent the continuous variation in the material properties of the nano-beam along its axis. The FG nanobeam is modeled using the Euler–Bernoulli beam theory, which incorporates the nonlocal theory of elasticity given by Eringen. The Rayleigh–Ritz method with Chebyshev polynomials is utilized to analyze the nondimensional frequency parameters of the nanobeam, providing insights into its dynamic behavior. Our results align well with existing literature in special cases. The nondimensional frequency parameters are obtained for the clamped-free boundary condition with different values of nondimensional small-scale parameters (e_0a/L), slenderness ratios (L/h) and power-law exponents (k). The convergence results are obtained. This study provides insights into the free vibration characteristics of axially FG cantilever nanobeam and demonstrates the effectiveness of the Rayleigh–Ritz method in analyzing such systems.

5.1 Introduction

Functionally graded (FG) materials offer unique advantages over traditional materials, making them ideal for demanding applications like aerospace structures, fusion reactors, and micromechanical systems. By gradually changing their properties, FG materials overcome the limitations in stiffness, stress concentration, and environmental resistance. Nanotechnology plays a crucial role in fabricating these advanced materials, enabling the development of revolutionary properties and enhanced functionalities at the nanoscale. Nanobeams, with their diverse

doi:10.1088/978-0-7503-6024-1ch5

applications in micromechanical systems, are a prime example of the potential of nanotechnology. Nanobeam structures, like nanowires, nanoprobes, and nano-actuators, are widely used in various systems and devices.

Axially FG nanobeams, with their smooth property variations along the axial direction of the beam, are particularly interesting for nanoelectromechanical systems. While previous studies have focused on thickness-wise property variations in FG materials, axial variations offer exciting possibilities at the nanoscale [1, 6]. Researchers have explored specific gradations like exponential, linear, and hyperbolic tangent, paving the way for further advancements in this promising field.

Alshorbagy *et al* [1] investigated the dynamic behavior of beams with varying material properties along their thickness. The material gradation followed a power-law and could occur either in the axial or transverse direction. They employed the finite element method to discretise the mathematical model and obtain solutions to the governing equation of motion. Eltaher *et al* [2] used a finite element method based on the nonlocal continuum model to analyze the free vibration characteristics of FG nanobeams. This model incorporates the size-dependent effects in FG nanobeams through the differential constitutive model of Eringen. Pradhan and Chakraverty [3] employed the Rayleigh–Ritz method to analyze the free vibrations of FG material beams. They compared the results obtained from both classical and first-order shear deformation beam theories. Chakraverty and Behera [4] utilized the Rayleigh–Ritz method to study the free vibration of nonuniform nanobeams and examined how the different material parameters affected the frequency parameters. Simsek *et al* [5] presented a new beam model that accounts for size-dependent effects to analyze the nonlinear free vibration characteristics of immovable-end FG nanobeams. They introduced a new reference surface in the formulation to eliminate the coupling between stretching and bending that arises due to the beam's thickness variations in material properties. Focusing on the buckling behavior of beams incorporating material and geometric complexities, Robinson and Adali [6] explored the case of axially FG and nonuniform Timoshenko beams resting on a Winkler–Pasternak foundation. They considered the small-scale effect and used the Rayleigh–Ritz method to analyze buckling loads for Timoshenko and Euler–Bernoulli beams. The vibration behavior of axially FG nanorods and nanobeams with a changing nonlocal parameter was studied by Aydogdu *et al* [7]. The Ritz method utilizing algebraic polynomials and stress gradient elasticity theory was employed to consider the structural effects at small scales and variations in material properties. The buckling behavior of Euler–Bernoulli nanobeams was investigated by Jena and Chakraverty [8] using the nonlocal elasticity theory given by Eringen. Using the Hermite–Ritz method and Navier's methodology, Jena *et al* [9] examined the vibration properties of a FG porous nanobeam placed in an elastic substrate. Bian *et al* [10] presented a finite element framework incorporating a stress-driven two-phase nonlocal integral model. This framework enables the analysis of mechanical responses in axially FG nanobeams under diverse boundary conditions. Chakraverty *et al* [11] examined the static and dynamic behaviors of structures made with FG materials. It covers various analyses like bending, buckling, and vibration, and presents advanced modeling techniques for static and dynamic problems.

This study investigates the free vibrations of an FG cantilever nanobeam made of steel and alumina (Al_2O_3). The special feature of this beam is that the material characteristics of the nanobeam vary continuously along its axial direction and are described by a power-law rule presented by Alshorbagy *et al* [1]. FG nanobeam is analyzed using the Euler–Bernoulli framework for beams alongside Eringen's theory, which accounts for nonlocal effects in elasticity. The Rayleigh–Ritz method with Chebyshev polynomials is used to find the nondimensional frequency parameters of the beam and understand its dynamic behavior.

5.2 FG Euler–Bernoulli cantilever beam

A FG cantilever beam is positioned in a cartesian coordinate system, with the origin denoted as $O(x, y, z)$. The beam possesses a rectangular-shaped cross-section and is characterized by the dimensions of length L, width b, and thickness h, as illustrated in figure 5.1.

The FG beam's material properties, specifically Young's modulus (E) and mass density (ρ), exhibit continuous variation along the beam axis as depicted in figure 5.2. This variation can be represented using a power-law expression [1] defined by

$$\mathcal{P}(x) = (\mathcal{P}_l - \mathcal{P}_r)\left(1 - \frac{x}{L}\right)^k + \mathcal{P}_r, \tag{5.1}$$

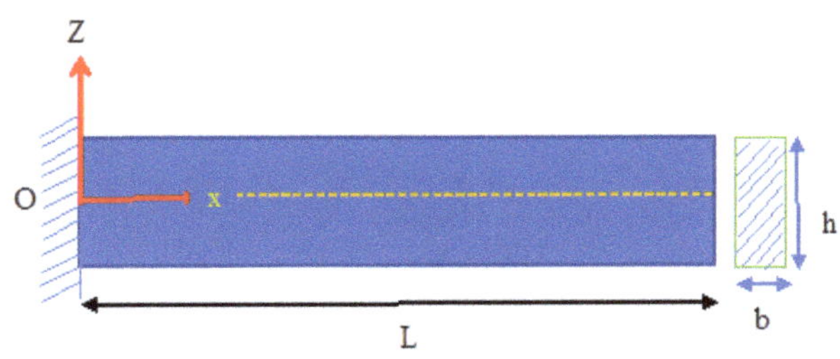

Figure 5.1. A typical FG cantilever beam.

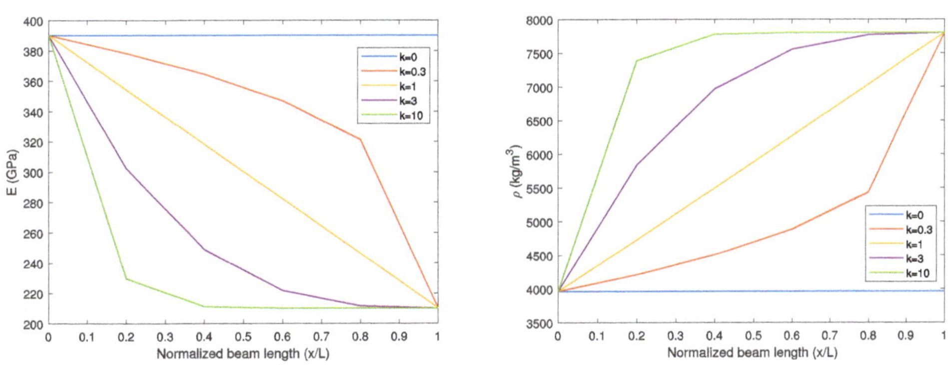

Figure 5.2. Power-law variation of E and ρ in the FG nanobeam along the beam's axes.

where the beam's effective material property at any given position, $\mathcal{P}(x)$, is determined by the characteristics of its left and right sides, represented by $\mathcal{P}_l$ and $\mathcal{P}_r$, respectively. k is the power-law exponent that is nonnegative.

The Young's modulus (E) and mass density (ρ) of the FG beam can be obtained from equation (5.1) and written as

$$\begin{aligned} E(x) &= (E_l - E_r)\left(1 - \frac{x}{L}\right)^k + E_r \\ \rho(x) &= (\rho_l - \rho_r)\left(1 - \frac{x}{L}\right)^k + \rho_r \end{aligned} \tag{5.2}$$

5.3 Theory and mathematical formulation

Any point in the beam's axial displacement (u) and transverse displacement (w), as determined by the Euler–Bernoulli beam theory, are represented as [12, 13]

$$\begin{bmatrix} u(x, z, t) \\ w(x, z, t) \end{bmatrix} = \begin{bmatrix} u_0(x, t) \\ w_0(x, t) \end{bmatrix} - z \begin{bmatrix} \dfrac{\partial w_0(x, t)}{\partial x} \\ 0 \end{bmatrix} \tag{5.3}$$

where t denotes time, $u_0(x, t)$ and $w_0(x, t)$ represent a point's axial and transverse displacements on the mid-plane.

In the context of the Euler–Bernoulli beam theory, assuming small deformations, only one strain component, ε_{xx}, is nonzero, and all other strains are zero. The strain ε_{xx} is given by [1]

$$\varepsilon_{xx} = \frac{\partial}{\partial x} u(x, z, t) = \frac{\partial}{\partial x}(u_0(x, t)) - z\frac{\partial^2}{\partial x^2}(w_0(x, t)) \tag{5.4}$$

Assuming the FG beam's material to be in compliance with Hooke's law, the constitutive equation for strain and stress is as follows

$$\begin{aligned} \sigma_{xx} &= E(x)\varepsilon_{xx} \\ &= E(x)\frac{\partial}{\partial x}(u_0(x, t)) - zE(x)\frac{\partial^2}{\partial x^2}(w_0(x, t)) \end{aligned} \tag{5.5}$$

where σ_{xx} is the normal stress.

The force resultant N is given by

$$N = \int_{-\frac{h}{2}}^{\frac{h}{2}} \sigma_{xx}(z)dz = D_0\frac{\partial u_0}{\partial x} - D_1\frac{\partial^2 w_0}{\partial x^2} \tag{5.6}$$

The moment resultant M is given by

$$M = \int_{-\frac{h}{2}}^{\frac{h}{2}} z\sigma_{xx}(z)dz = D_1\frac{\partial u_0}{\partial x} - D_2\frac{\partial^2 w_0}{\partial x^2} \tag{5.7}$$

where D_0, D_1 and D_2 represent extensional, coupling, and bending rigidities, which can be expressed as

$$(D_0, D_1, D_2) = \int_{-\frac{h}{2}}^{\frac{h}{2}} E(x)(1, z, z^2)dz \tag{5.8}$$

The equations for the strain energy (S) and kinetic energy (T) of a beam with cross-sectional area A and length L, at any instant are expressed as [3, 14]

$$S = \frac{1}{2}\int_0^L \int_A \sigma_{xx}\varepsilon_{xx}\, dAdx \tag{5.9}$$

$$T = \frac{1}{2}\int_0^L \int_A \rho(x)\left(\left(\frac{\partial u}{\partial t}\right)^2 + \left(\frac{\partial w}{\partial t}\right)^2\right)dAdx \tag{5.10}$$

Equations (5.9) and (5.10) can be rewritten using equations (5.3)–(5.8) and expressed as [15]

$$\begin{aligned} S &= \frac{1}{2}\int_0^L \left(N\left(\frac{\partial u_0}{\partial x}\right) - M\left(\frac{\partial^2 w_0}{\partial x^2}\right)\right)dx \\ &= \frac{1}{2}\int_0^L \left(D_0\left(\frac{\partial u_0}{\partial x}\right)^2 - 2D_1\left(\frac{\partial^2 w_0}{\partial x^2}\right)\left(\frac{\partial u_0}{\partial x}\right) + D_2\left(\frac{\partial^2 w_0}{\partial x^2}\right)^2\right)dx \end{aligned} \tag{5.11}$$

$$T = \frac{1}{2}\int_0^L \left(I_0\left(\frac{\partial u_0}{\partial t}\right)^2 + I_0\left(\frac{\partial w_0}{\partial t}\right)^2 - 2I_1\left(\frac{\partial u_0}{\partial t}\right)\left(\frac{\partial^2 w_0}{\partial x\partial t}\right) + I_2\left(\frac{\partial^2 w_0}{\partial x\partial t}\right)^2\right)dx \tag{5.12}$$

The mass moment of inertia appearing in equation (5.12) are expressed as follows

$$(I_0, I_1, I_2) = \int_{-\frac{h}{2}}^{\frac{h}{2}} \rho(x)(1, z, z^2)dz \tag{5.13}$$

The equilibrium equations for free vibration analysis can be obtained using the Euler–Lagrange equation, which is expressed as [2, 15]

$$\frac{\partial N}{\partial x} = I_0\frac{\partial^2 u_0}{\partial t^2} - I_1\frac{\partial^3 w_0}{\partial x\partial t^2} \tag{5.14}$$

$$\frac{\partial^2 M}{\partial x^2} = I_0\frac{\partial^2 w_0}{\partial t^2} + I_1\frac{\partial^3 u_0}{\partial x\partial t^2} - I_2\frac{\partial^4 w_0}{\partial x^2\partial t^2} \tag{5.15}$$

5.4 Nonlocal FG Euler–Bernoulli nanobeam

The Eringen nonlocal constitutive relations can be written in a differentiable form by [16, 17]

$$(1 - (e_0a)^2\nabla^2)\sigma_{ij} = C_{ijkl}\varepsilon_{kl} \tag{5.16}$$

where e_0 represents a constant to modulate the model according to the robustness of the results obtained from experiments or from other models, a represents the intrinsic characteristic length (i.e. the lattice parameter, the granular size, or the molecular diameters), σ_{ij} is the nonlocal stress tensor, C_{ijkl} is the fourth-order elasticity tensor, and ε_{kl} is the strain tensor.

Equation (5.16) can be rephrased as follows for Euler–Bernoulli nonlocal FG nanobeam:

$$\sigma_{xx} - (e_0 a)^2 \frac{\partial^2 \sigma_{xx}}{\partial x^2} = E(x)\varepsilon_{xx} \tag{5.17}$$

The axial force-strain relation can be derived by integrating equation (5.17) over the cross-sectional area of the FG nanobeam, which can be written as

$$N - (e_0 a)^2 \frac{\partial^2 N}{\partial x^2} = D_0 \frac{\partial u_0}{\partial x} - D_1 \frac{\partial^2 w_0}{\partial x^2} \tag{5.18}$$

The moment-curvature relation can be derived by multiplying equation (5.17) by z and integrating it across the beam's cross-sectional area. This relation can be stated as

$$M - (e_0 a)^2 \frac{\partial^2 M}{\partial x^2} = D_1 \frac{\partial u_0}{\partial x} - D_2 \frac{\partial^2 w_0}{\partial x^2} \tag{5.19}$$

Equation (5.14) can be differentiated with respect to x, and then the obtained result can be substituted into equation (5.18) to get

$$N = D_0 \frac{\partial u_0}{\partial x} - D_1 \frac{\partial^2 w_0}{\partial x^2} + (e_0 a)^2 \left(I_0 \frac{\partial^3 u_0}{\partial x \partial t^2} - I_1 \frac{\partial^4 w_0}{\partial x^2 \partial t^2} \right) \tag{5.20}$$

The second derivative of M from equation (5.15) can be substituted into equation (5.19) to obtain

$$M = D_1 \frac{\partial u_0}{\partial x} - D_2 \frac{\partial^2 w_0}{\partial x^2} + (e_0 a)^2 \left(I_0 \frac{\partial^2 w_0}{\partial t^2} + I_1 \frac{\partial^3 u_0}{\partial x \partial t^2} - I_2 \frac{\partial^4 w_0}{\partial x^2 \partial t^2} \right) \tag{5.21}$$

5.5 Rayleigh–Ritz formulation

For an axially FG Euler–Bernoulli nanobeam, the strain and kinetic energies have the following form [15]

$$S = \frac{1}{2} \int_0^L \left(\begin{array}{l} \left(D_0 \left(\frac{\partial u_0}{\partial x} \right)^2 - D_1 \frac{\partial^2 w_0}{\partial x^2} \frac{\partial u_0}{\partial x} + (e_0 a)^2 \left(I_0 \frac{\partial^3 u_0}{\partial x \partial t^2} \frac{\partial u_0}{\partial x} - I_1 \frac{\partial^4 w_0}{\partial x^2 \partial t^2} \frac{\partial u_0}{\partial x} \right) \right) \\ - \left(\begin{array}{l} D_1 \frac{\partial u_0}{\partial x} \frac{\partial^2 w_0}{\partial x^2} - D_2 \left(\frac{\partial^2 w_0}{\partial x^2} \right)^2 \\ + (e_0 a)^2 \left(I_0 \frac{\partial^2 w_0}{\partial t^2} \frac{\partial^2 w_0}{\partial x^2} + I_1 \frac{\partial^3 u_0}{\partial x \partial t^2} \frac{\partial^2 w_0}{\partial x^2} - I_2 \frac{\partial^4 w_0}{\partial x^2 \partial t^2} \frac{\partial^2 w_0}{\partial x^2} \right) \end{array} \right) \end{array} \right) dx \tag{5.22}$$

$$T = \frac{1}{2}\int_0^L \left(I_0\left(\left(\frac{\partial u_0}{\partial t}\right)^2 + \left(\frac{\partial w_0}{\partial t}\right)^2\right) - 2I_1\left(\frac{\partial u_0}{\partial t}\right)\left(\frac{\partial^2 w_0}{\partial x \partial t}\right) + I_2\left(\frac{\partial^2 w_0}{\partial x \partial t}\right)^2 \right) dx \quad (5.23)$$

In Euler–Bernoulli beam theory, $u_0(x, t)$ and $w_0(x, t)$ can be expressed as [19]

$$\begin{aligned} u_0(x, t) &= U_0(x)\cos(\omega t), \\ w_0(x, t) &= W_0(x)\cos(\omega t) \end{aligned} \quad (5.24)$$

where the trigonometric terms indicate the free vibration's harmonic type response, ω denotes the natural frequency, $U_0(x)$ denotes the amplitudes for the axial displacement and $W_0(x)$ denotes the amplitudes for the transverse displacement of the free vibration of the FG nanobeam.

The expressions (5.24) can be substituted into equations (5.22) and (5.23) to obtain the maximum strain and kinetic energies, which can be expressed as

$$S_{max} = \frac{1}{2}\int_0^L \left(\begin{array}{l} \left(D_0\left(\frac{dU_0}{dx}\right)^2 - D_1\frac{d^2W_0}{dx^2}\frac{dU_0}{dx} - \omega^2(e_0a)^2\left(I_0\left(\frac{dU_0}{dx}\right)^2 - I_1\frac{d^2W_0}{dx^2}\frac{dU_0}{dx}\right)\right) \\ -\left(\begin{array}{l} D_1\frac{dU_0}{dx}\frac{d^2W_0}{dx^2} - D_2\left(\frac{d^2W_0}{dx^2}\right)^2 \\ -\omega^2(e_0a)^2\left(I_0 W_0\frac{d^2W_0}{dx^2} + I_1\frac{dU_0}{dx}\frac{d^2W_0}{dx^2} - I_2\left(\frac{d^2W_0}{dx^2}\right)^2\right) \end{array} \right) \end{array} \right) dx \quad (5.25)$$

$$T_{max} = -\frac{\omega^2}{2}\int_0^L \left(I_0\left(U_0^2 + W_0^2\right) - 2I_1 U_0\frac{dW_0}{dx} + I_2\left(\frac{dW_0}{dx}\right)^2 \right) dx \quad (5.26)$$

For the Rayleigh–Ritz method, the displacement functions can be expressed by

$$\begin{aligned} U_0(x) &= x^p(L - x)^q\sum_{i=1}^{N} c_i\phi_i \\ W_0(x) &= x^p(L - x)^q\sum_{j=1}^{N} d_j\psi_j \end{aligned} \quad (5.27)$$

where N denotes the number of polynomials considered in the displacement functions. The unknown constant coefficients c_i and d_j are to be determined. The Chebyshev polynomials are considered as shape functions ϕ_i and ψ_j for the analysis. For cantilever beam, $p = 2$ and $q = 0$ [3].

5.5.1 Chebyshev polynomials

Chebyshev polynomials of the first kind, represented as $T_n(x)$, are a sequence of orthogonal polynomials. The first two terms with recurrence relation of this sequence are represented as [18]

$$
\begin{aligned}
T_0(x) &= 1 \\
T_1(x) &= x \\
&\vdots \\
T_n(x) &= 2xT_{n-1}(x) - T_{n-2}(x), \quad n = 2, 3, 4, \cdots
\end{aligned} \tag{5.28}
$$

The utilization of Chebyshev polynomials presents significant advantages, primarily due to their orthogonal properties. Moreover, when n exceeds 10, the system effectively avoids ill-conditioning problems [18].

In this analysis, we have considered $\phi_i = T_{i-1}(x)$ and $\psi_j = T_{j-1}(x)$ where $i, j = 1, 2, \cdots, N$.

To establish the governing equation for the free vibration in FG nanobeams, we first obtain the Rayleigh quotient by equating S_{max} with T_{max}. Then the Rayleigh quotient can be partially differentiated with respect to each constant coefficient (c_i and d_j; $i, j = 1, 2, \cdots, N$) to produce a set of N equations, which can subsequently be formulated as a generalized eigenvalue problem, as demonstrated below

$$([K] - \lambda^2[M])\{ \Delta \} = 0 \tag{5.29}$$

where $[K]$ represent the stiffness matrix and $[M]$ represent the inertia matrix. The eigenvalues λ represent the nondimensional frequency parameters for the free vibration problem of the axially FG nanobeam, and $\{ \Delta \}$ denotes the column vector containing the unknown constant coefficients.

5.6 Results and discussion

The initial five nondimensional frequency parameters for the free vibration of axially FG cantilever nanobeams are tabulated and analyzed in this section. The effects of slenderness ratio (L/h), nondimensional small-scale parameter (e_0a/L), and power-law exponent (k) on these parameters are investigated. Also, the convergence results and model validation are presented to ensure the accuracy and reliability of the findings.

The FG nanobeam is constructed using a steel–alumina (Al_2O_3) composite. The material properties of this composite vary continuously along the beam's axial direction, following a power-law distribution. The right side of the beam is composed entirely of steel, while the left side is composed entirely of alumina. The material properties of both steel and alumina are provided in table 5.1 [1, 2].

Table 5.1. Material properties of FG nanobeams constituents [1, 2].

	Properties	
	E (GPa)	ρ (kg m^{-3})
Steel	210	7800
Alumina (Al_2O_3)	390	3960

The nondimensional frequency parameters used here are defined as

$$\lambda = \omega L^2 \sqrt{\frac{\rho_l A}{E_l I}} \tag{5.30}$$

where A is the cross-sectional area, and I is the moment of inertia and calculated as $I = \frac{bh^3}{12}$ for a beam with rectangular cross-section of width b, thickness h.

5.6.1 Convergence study

The numerical convergence behaviors of the initial five nondimensional frequency parameters are analyzed in tabular formats for four different parameter combinations:

1. $\frac{L}{h} = 5$, $\frac{e_0 a}{L} = 0$, $k = 0$
2. $\frac{L}{h} = 5$, $\frac{e_0 a}{L} = 0$, $k = 10$
3. $\frac{L}{h} = 5$, $\frac{e_0 a}{L} = 0.7$, $k = 0$
4. $\frac{L}{h} = 5$, $\frac{e_0 a}{L} = 0.7$, $k = 10$

and presented in tables 5.2–5.5.

The results show that increasing the number of polynomials used in the Rayleigh–Ritz method significantly improves the accuracy of nondimensional frequency parameters for axially FG cantilever nanobeams. The convergence pattern of axially FG nanobeams can be examined for different slenderness ratios, nondimensional small-scale parameters, and power-law exponents. To ensure convergence while maintaining computational efficiency, $N = 11$ was chosen for subsequent analyses.

Table 5.2. Convergence analysis of the initial five nondimensional frequency parameters for $L/h = 5$, $e_0 a/L = 0$ and $k = 0$.

N	λ_1	λ_2	λ_3	λ_4	λ_5
2	3.5047	30.6199	31.8411	134.6937	—
3	3.4901	21.0781	28.9789	87.3820	93.8918
4	3.4891	21.0378	28.3320	56.2124	84.4530
5	3.4891	20.9302	27.9814	56.2124	83.8385
6	3.4891	20.9299	27.7725	55.0441	83.2941
7	3.4891	20.9297	27.6382	55.0411	82.8998
8	3.4891	20.9297	27.5467	55.0279	82.6318
9	3.4891	20.9297	27.4816	55.0279	82.4401
10	3.4891	20.9297	27.4337	55.0279	82.2985
11	3.4891	20.9297	27.4007	55.0279	82.1999

Table 5.3. Convergence analysis of the initial five nondimensional frequency parameters for $L/h = 5$, $e_0a/L = 0$ and $k = 10$.

N	λ_1	λ_2	λ_3	λ_4	λ_5
2	2.0208	16.1959	16.8752	70.6312	—
3	2.0178	11.8418	15.5166	46.5833	49.6395
4	2.0023	11.8073	15.3274	31.8517	45.6389
5	2.0008	11.6128	15.2474	31.3949	45.5756
6	2.0007	11.6113	15.2027	30.2656	45.4811
7	2.0007	11.6113	15.1716	30.2605	45.3927
8	2.0007	11.6112	15.1483	30.2454	45.3233
9	2.0007	11.6112	15.1306	30.2452	45.2689
10	2.0007	11.6111	15.1172	30.2452	45.2277
11	2.0007	11.6111	15.1067	30.2452	45.1958

Table 5.4. Convergence analysis of the initial five nondimensional frequency parameters for $L/h = 5$, $e_0a/L = 0.7$ and $k = 0$.

N	λ_1	λ_2	λ_3	λ_4	λ_5
2	3.2573	—	—	—	—
3	3.2570	7.3221	—	—	—
4	3.2480	7.3221	12.0835	—	—
5	3.2480	7.2197	12.0835	15.1049	—
6	3.2480	7.2197	11.5223	15.1049	17.9941
7	3.2480	7.2192	11.5223	14.1862	17.9941
8	3.2480	7.2192	11.5079	14.1862	16.6191
9	3.2480	7.2192	11.5079	14.1408	16.6191
10	3.2480	7.2192	11.5077	14.1408	16.4930
11	3.2480	7.2192	11.5077	14.1400	16.4926

5.6.2 Validation

To ensure the accuracy of our analysis, we compared the initial five nondimensional frequency parameters of the axially FG cantilever nanobeam with those presented in [2] for slenderness ratios of $L/h = 100$ and a power-law exponent of $k = 0$ with different nondimensional small-scale parameter e_0a/L. This comparison, shown in table 5.6, reveals that the numerical results obtained in our present analysis for the axially FG cantilever nanobeam closely align with those presented in [2], validating the present study's approach for analyzing this type of nanobeam.

Table 5.5. Convergence analysis of the initial five nondimensional frequency parameters for $L/h = 5$, $e_0a/L = 0.7$ and $k = 10$.

N	λ_1	λ_2	λ_3	λ_4	λ_5
2	1.8817	—	—	—	—
3	1.8610	4.1056	—	—	—
4	1.8332	4.0420	7.0317	—	—
5	1.8324	3.9255	6.7320	9.1650	—
6	1.8324	3.9254	6.2999	8.3674	11.6557
7	1.8324	3.9244	6.2829	7.7984	9.8861
8	1.8323	3.9243	6.2594	7.7227	9.3575
9	1.8323	3.9243	6.2594	7.6676	9.0704
10	1.8323	3.9243	6.2592	7.6657	8.9818
11	1.8323	3.9243	6.2592	7.6635	8.9555

5.6.3 Parametric results

The initial five nondimensional frequency parameters for the axially FG cantilever nanobeam, made from steel and alumina, are presented in tables 5.7–5.10 for various nondimensional small-scale parameters ($e_0a/L = 0, 0.1, 0.2, 0.4$) and power-law exponents ($k = 0, 0.3, 1, 3, 10$) across different slenderness ratios ($L/h = 10, 20, 50, 100$).

It is noted that, as the power-law exponents increase from 0 to 10, the initial five nondimensional frequency parameters exhibit a decreasing trend. This is due to the transition of the beam from alumina to steel, as alumina has a higher Young's modulus and lower mass density, and steel has a lower Young's modulus and higher mass density. The initial nondimensional frequency parameter increases with the increase of nondimensional small-scale parameters from 0 to 0.4, while the second to fifth nondimensional frequency parameters decrease over the same range. Notably, an increase in slenderness ratios from 10 to 100 increases the values of the nondimensional frequency parameters.

5.6.4 Graphical results

The effect of the power-law exponent and nondimensional small-scale parameters on the initial five nondimensional frequency parameters is presented in figure 5.3, with a specific slenderness ratio of $L/h = 100$.

Again, it can be observed that the first nondimensional frequency parameter increases as the nondimensional small-scale parameter goes from 0 to 0.4. In contrast, the second to fifth parameters decrease within the same range. The nondimensional small-scale parameter has a more substantial influence on the higher-order nondimensional frequency parameters (λ_3, λ_4, λ_5) than the lower ones (λ_1, λ_2), as shown in figure 5.3. The rate of decrease for all five nondimensional

Table 5.6. Comparison of the initial five nondimensional frequency parameters with Eltaher [2] ($L/h = 100$, $k = 0$).

e_0a/L	λ_1		λ_2		λ_3		λ_4		λ_5	
	Present	[2]	Present	[2]	Present	[2]	Present	[2]	Present	[2]
0	3.5159	3.5167	22.0315	22.0378	61.6774	61.7173	120.8300	120.9743	199.6956	200.0507
10^{-1}	3.5306	3.5292	20.6815	20.6817	51.0424	51.0694	85.6431	85.6894	121.2693	121.3309
$\sqrt{2}\cdot 10^{-1}$	3.5445	3.5461	19.5270	19.5105	44.5187	44.5603	70.0177	70.0139	95.1228	95.1715
$\sqrt{3}\cdot 10^{-1}$	3.5575	3.5632	18.5282	18.4860	40.0129	40.0946	60.7017	60.6293	80.9234	81.0297

Table 5.7. Variation of nondimensional frequency parameters for different values of e_0a/L and k with fixed $L/h = 10$.

e_0a/L	k	λ_1	λ_2	λ_3	λ_4	λ_5
	0	3.5092	21.7425	54.8014	59.8012	114.2898
	0.3	2.8908	18.4475	45.8806	50.8841	97.1859
0	1	2.4967	15.4553	38.6494	42.4990	81.1168
	3	2.2282	13.0669	33.2985	36.0314	68.9613
	10	2.0130	12.0789	30.2134	32.9347	62.7965
	0	3.5237	20.4069	49.4850	55.4932	81.0164
	0.3	2.9018	17.3178	42.0969	46.3793	68.8272
0.1	1	2.5070	14.4878	35.0428	39.0464	57.1676
	3	2.2387	12.2254	29.6188	33.6849	48.4452
	10	2.0223	11.3073	27.1216	30.5909	44.2257
	0	3.5620	17.4228	35.5504	51.3896	57.7609
	0.3	2.9288	14.7079	29.9952	43.3396	47.9468
0.2	1	2.5319	12.2656	24.9483	35.9989	40.3433
	3	2.2655	10.3528	21.1270	30.5573	34.8952
	10	2.0465	9.5976	19.3860	27.9347	31.7878
	0	3.6233	11.8755	20.9400	27.9457	34.8136
	0.3	2.9391	9.8437	17.5024	23.4672	29.2295
0.4	1	2.5325	8.1649	14.5961	19.5147	24.3108
	3	2.2877	6.9475	12.4065	16.5821	20.7405
	10	2.0796	6.4813	11.3867	15.1603	18.9419

Table 5.8. Variation of nondimensional frequency parameters for different values of e_0a/L and k with fixed $L/h = 20$.

e_0a/L	k	λ_1	λ_2	λ_3	λ_4	λ_5
	0	3.5143	21.9604	61.2063	109.6028	119.1414
	0.3	2.8948	18.6482	52.1755	91.7612	101.6039
0	1	2.5003	15.6331	43.6247	77.2633	84.9276
	3	2.2316	13.2102	36.9464	66.5970	72.0915
	10	2.0161	12.2046	33.7299	60.4269	65.5313
	0	3.5290	20.6140	50.6518	84.4509	110.9863
	0.3	2.9059	17.5039	43.1243	71.8141	92.7585
0.1	1	2.5107	14.6517	35.9348	59.7512	78.0929
	3	2.2422	12.3585	30.3515	50.5861	67.3697
	10	2.0255	11.4249	27.7642	46.1059	61.1817
	0	3.5678	17.5985	36.3859	53.5791	70.0784
	0.3	2.9331	14.8564	30.7033	45.2296	59.1658

(*Continued*)

Table 5.8. (*Continued*)

e_0a/L	k	λ_1	λ_2	λ_3	λ_4	λ_5
0.2	1	2.5357	12.3941	25.5609	37.6406	49.2612
	3	2.2692	10.4601	21.6354	31.9122	41.8244
	10	2.0500	9.6943	19.8357	29.1198	38.1319
	0	3.6304	11.9908	21.4576	29.1653	37.2074
	0.3	2.9440	9.9363	17.9437	24.5272	31.3332
0.4	1	2.5366	8.2432	14.9809	20.4408	26.1436
	3	2.2918	7.0141	12.7221	17.3407	22.2446
	10	2.0837	6.5430	11.6633	15.8188	20.2484

Table 5.9. Variation of nondimensional frequency parameters for different values of e_0a/L and k with fixed $L/h = 50$.

e_0a/L	k	λ_1	λ_2	λ_3	λ_4	λ_5
	0	3.5157	22.0226	61.6179	120.6150	199.1296
	0.3	2.8959	18.7055	52.5547	102.9518	169.9792
0	1	2.5013	15.6839	43.9560	86.0941	142.1575
	3	2.2326	13.2511	37.2154	73.0477	120.8051
	10	2.0170	12.2405	33.9631	66.3638	109.5649
	0	3.5304	20.6730	50.9931	85.4914	120.9285
	0.3	2.9070	17.5570	43.4246	72.7181	102.7063
0.1	1	2.5117	14.6984	36.1958	60.5338	85.4728
	3	2.2432	12.3966	30.5660	51.2350	72.4864
	10	2.0264	11.4584	27.9521	46.6750	65.9664
	0	3.5694	17.6486	36.6303	54.2437	71.4700
	0.3	2.9343	14.8987	30.9103	45.8019	60.3729
0.2	1	2.5368	12.4306	25.7401	38.1381	50.3282
	3	2.2703	10.4907	21.7841	32.3235	42.7294
	10	2.0509	9.7219	19.9671	29.4788	38.8890
	0	3.6324	12.0236	21.6095	29.5364	37.9701
	0.3	2.9454	9.9626	18.0728	24.8484	31.9976
0.4	1	2.5377	8.2655	15.0937	20.7218	26.7410
	3	2.2930	7.0330	12.8146	17.5711	22.7188
	10	2.0848	6.5605	11.7443	16.0187	20.6640

frequency parameters is much faster within the power-law exponent range of 0–3 compared to 3–10.

The effect of nondimensional small-scale parameters on mode numbers for various power-law exponents ($k = 0$, 0.3, 1, 3, 10) is presented in figure 5.4 at a fixed slenderness ratio of $L/h = 50$.

Table 5.10. Variation of nondimensional frequency parameters for different values of e_0a/L and k with fixed $L/h = 100$.

e_0a/L	k	λ_1	λ_2	λ_3	λ_4	λ_5
0	0	3.5159	22.0315	61.6774	120.8300	199.6956
	0.3	2.8961	18.7137	52.6095	103.1487	170.4943
	1	2.5014	15.6912	44.0039	86.2646	142.6012
	3	2.2327	13.2570	37.2543	73.1873	121.1731
	10	2.0171	12.2456	33.9968	66.4853	109.8852
0.1	0	3.5306	20.6815	51.0424	85.6431	121.2693
	0.3	2.9072	17.5646	43.4680	72.8499	103.0046
	1	2.5118	14.7051	36.2335	60.6479	85.7427
	3	2.2433	12.4020	30.5970	51.3298	72.7072
	10	2.0265	11.4632	27.9792	46.7579	66.1531
0.2	0	3.5696	17.6558	36.6656	54.3407	71.6765
	0.3	2.9345	14.9047	30.9402	45.8853	60.5551
	1	2.5369	12.4359	25.7659	38.2107	50.4825
	3	2.2704	10.4951	21.8056	32.3831	42.8275
	10	2.0511	9.7259	19.9861	29.5312	39.0007
0.4	0	3.6326	12.0284	21.6315	29.5906	38.0837
	0.3	2.9456	9.9664	18.0915	24.8953	32.0975
	1	2.5379	8.2687	15.1100	20.7628	26.8205
	3	2.2931	7.0358	12.8280	17.6048	22.7962
	10	2.0850	6.5631	11.7560	16.0480	20.7255

As the mode number increases (for a constant nondimensional small-scale parameter), the values of nondimensional frequency parameters also increase. Higher modes of vibration correspond to higher frequencies. Conversely, increasing the nondimensional small-scale parameter (for any fixed mode number) decreases the values of the nondimensional frequency parameters. The same effect is obtained for a range of power-law exponents ($k = 0$ to 10), as shown in figure 5.4. This indicates that the observed effect of the nondimensional small-scale parameter on the mode number is independent of the material graduation within the investigated range.

5.7 Conclusions

This chapter analyses the free vibration characteristics of an axially FG steel–alumina cantilever nanobeam using the Euler–Bernoulli beam theory, which incorporates the nonlocal theory of elasticity given by Eringen. Material characteristics of the beam follow a power-law form and vary continuously along the beam's axes. The Rayleigh–Ritz method with Chebyshev polynomials is used to determine the nondimensional frequency parameters. Convergence studies and comparisons with existing literature

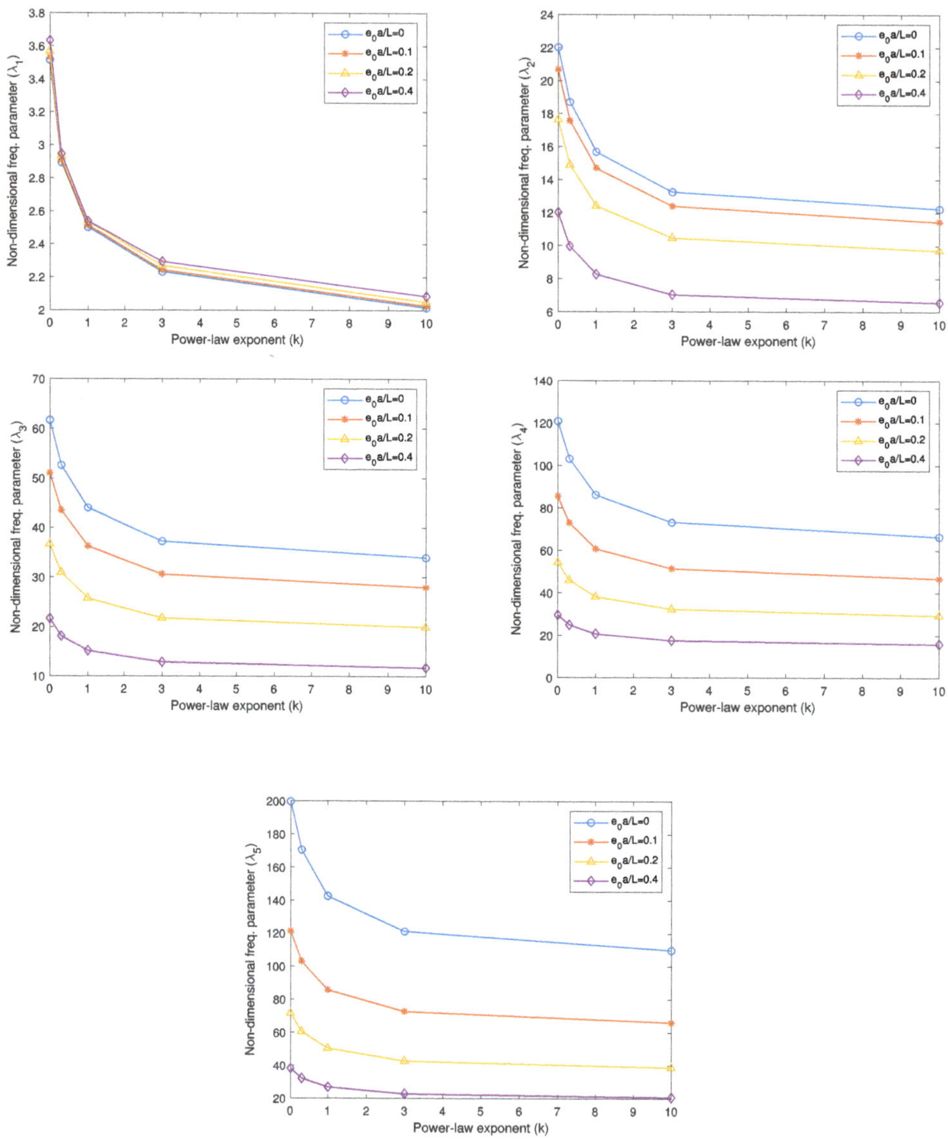

Figure 5.3. Variation of the initial five nondimensional frequency parameters with power-law exponent for different nondimensional small-scale parameters at $L/h = 100$.

confirm the accuracy of the results. The effects of the small-scale parameters, power-law exponents, and slenderness ratios on nondimensional frequency parameters are explored. The analysis draws the following important conclusions:

- The nonuniform distribution of material along a beam's axial direction results in varying stiffness across its length. This uneven stiffness affects the frequency parameters and mode shapes of the beam.

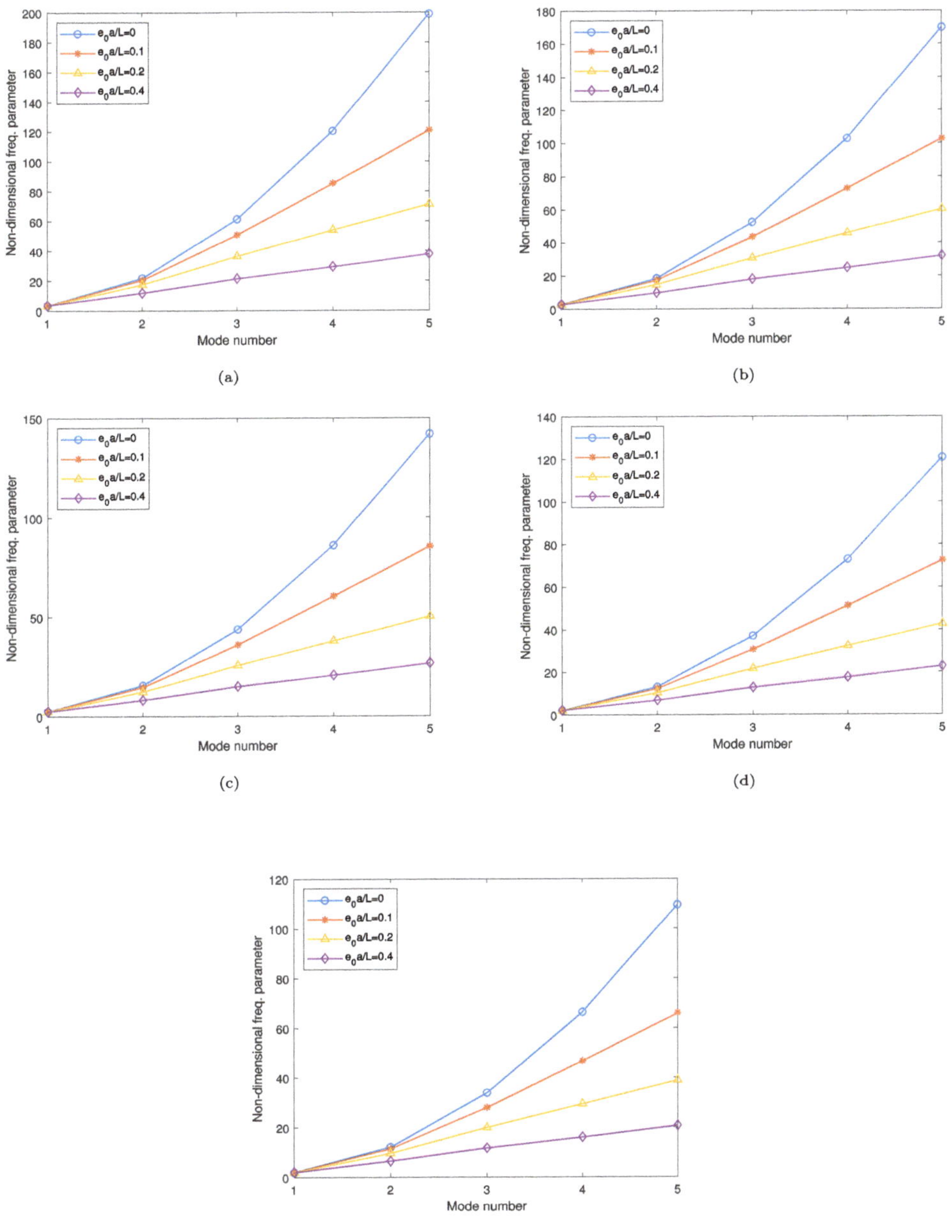

Figure 5.4. The effect of nondimensional small-scale parameter on mode number at $L/h = 50$ for (a) $k = 0$, (b) $k = 0.3$, (c) $k = 1$, (d) $k = 3$, (e) $k = 10$.

- The nondimensional frequency parameter decreases as the power-law exponent increases from 0 to 10. The rate of decrease is much faster within the range of 0 to 3 compared to 3 to 10.

- As the nondimensional small-scale parameter increases from 0 to 0.4, the first nondimensional frequency parameter increases, while the second to fifth decreases. Its impact on higher-order nondimensional frequency parameters is more significant than on lower-order ones.
- As the slenderness ratio increases from 10 to 100, the values of the nondimensional frequency parameters also significantly increase.
- For a fixed nondimensional small-scale parameter, the nondimensional frequency parameter increases as the mode number increases.
- For a fixed mode number, an increase in the nondimensional small-scale parameter consistently decreases the values of the nondimensional frequency parameters.
- The effect of the nondimensional small-scale parameter on the mode number is independent of the material graduation within the investigated power-law exponent range of $k = 0$ to 10.

References

[1] Alshorbagy A E, Eltaher M and Mahmoud F 2011 Free vibration characteristics of a functionally graded beam by finite element method *Appl. Math. Modelling* **35** 412–25

[2] Eltaher M, Emam S A and Mahmoud F 2012 Free vibration analysis of functionally graded size-dependent nanobeams *Appl. Math. Comput.* **218** 7406–20

[3] Pradhan K and Chakraverty S 2013 Free vibration of Euler and Timoshenko functionally graded beams by Rayleigh-Ritz method *Compos. Part B: Eng.* **51** 175–84

[4] Chakraverty S and Behera L 2015 Free vibration of non-uniform nanobeams using Rayleigh-Ritz method *Physica E: Low-dimensional Syst. Nanostruct.* **67** 38–46

[5] Şimşek M 2016 Nonlinear free vibration of a functionally graded nanobeam using nonlocal strain gradient theory and a novel Hamiltonian approach *Int. J. Eng. Sci.* **105** 12–27

[6] Robinson M T A and Adali S 2018 Buckling of nonuniform and axially functionally graded nonlocal Timoshenko nanobeams on Winkler-Pasternak foundation *Compos. Struct.* **206** 95–103

[7] Aydogdu M, Arda M and Filiz S 2018 Vibration of axially functionally graded nano rods and beams with a variable nonlocal parameter *Adv. Nano Res.* **6** 257–78

[8] Jena S K and Chakraverty S 2019 Differential quadrature and differential transformation methods in buckling analysis of nanobeams *Curved Layered Struct.* **6** 68–76

[9] Jena S, Chakraverty S, Malikan M and Sedighi H M 2020 Implementation of Hermite-Ritz method and Navier's technique for vibration of functionally graded porous nanobeam embedded in Winkler-Pasternak elastic foundation using bi-helmholtz nonlocal elasticity *J. Mech. Mater. Struct.* **15** 405–34

[10] Bian P L, Qing H and Yu T 2022 A new finite element method framework for axially functionally-graded nanobeam with stress-driven two-phase nonlocal integral model *Compos. Struct.* **295** 115769

[11] Chakraverty S, Jena S K and Civalek Ö (ed) 2023 *Functionally Graded Structures,* (Bristol: IOP Publishing) pp 2053–563 https://dx.doi.org/10.1088/978-0-7503-5301-4

[12] Şimşek M and Kocatürk T 2009 Free and forced vibration of a functionally graded beam subjected to a concentrated moving harmonic load *Compos. Struct.* **90** 465–73

[13] Chakraverty S 2008 *Vibration of Plates* (Boca Raton, FL: CRC Press) https://doi.org/10.1201/9781420053968

[14] Wang C, Reddy J N and Lee K 2000 *Shear Deformable Beams and Plates: Relationships with Classical Solutions* (Amsterdam: Elsevier) https://doi.org/10.1016/B978-0-08-043784-2.X5000-X

[15] Uzun B and Özgür Yaylı M 2019 Finite element model of functionally graded nanobeam for free vibration analysis *Int. J. Eng. Appl. Sci.* **11** 387–400

[16] Eringen A C 1983 On differential equations of nonlocal elasticity and solutions of screw dislocation and surface waves *J. Appl. Phys.* **54** 4703–10

[17] Eringen A A and Wegner J R 2003 Nonlocal continuum field theories *Appl. Mech. Rev.* **56** B20–2

[18] Jena S K, Harursampath D, Mahesh V and Ponnusami S A 2022 Comparing different polynomials-based shape functions in the Rayleigh–Ritz method for investigating dynamical characteristics of nanobeam *Polynomial Paradigms* (Bristol: IOP Publishing) pp 2053–563 https://doi.org/10.1088/2053-2563/ac9580CH007

[19] Gartia A K and Chakraverty S 2024 Free Vibration of Bi-Directional Functionally Graded Nanobeams Resting on Winkler–Pasternak Foundations *J. Vib. Engi. Technol.* **2024** 1–17

Chapter 6

Flexural vibration of rectangular functionally graded nanoplates

K K Pradhan and S Chakraverty

In this chapter, flexural vibration of functionally graded nanoplates based on classical plate theory in conjunction with Eringen's nonlocal elasticity theory has been investigated. The material properties in such a plate vary spatially across the thickness in power-law form. Numerical modeling is carried out by means of the Rayleigh–Ritz method, in which trial functions defining the transverse displacement are expressed as linear combinations of simple algebraic polynomials. The effect of different physical parameters on the eigenfrequencies has been addressed here after a valid comparison with available results.

6.1 Introduction

In recent years, nanosized structures have gained attention in the scientific community due to their extraordinary physical, chemical, mechanical, and electrical properties [1]. Performing precise experiments at such a scale is not only challenging, but also cumbersome and expensive [2]. It is rightly stated in [3] that the nanoparticles act as lubricant additives in lubricated systems and the mechanical properties of nanoparticles will influence the tribological properties helping to increase lubricating behavior. Anjomshoa [4] has applied Ritz functions in buckling analysis of embedded orthotropic circular and elliptic micro/nanoplates based on nonlocal elasticity theory. An analytical Hamiltonian-based model is proposed by Rong *et al* [5] for the dynamic analysis of rectangular nanoplates using Kirchoff plate theory and Eringen's nonlocal theory. Singh and Azam [6] have investigated free vibration and buckling behaviors of functionally graded plate supported on Winkler-Pasternak elastic foundation based on nonlocal elasticity theory. Due to the potential application of nanostructural elements, free vibration of functionally graded (FG) rectangular nanoplates subject to classical boundary conditions is investigated in this chapter based on the Rayleigh–Ritz method and the effect of physical parameters on natural frequencies can be observed here along with mode shapes in various cases.

doi:10.1088/978-0-7503-6024-1ch6

6.2 FG nanoplate

In this section, a FG rectangular nanoplate of respective length a, breadth b, and thickness h is considered which is subject to various sets of classical boundary conditions. The FG nanoplate here has a ceramic bottom surface and metal top surface. Material properties viz. Young's modulus and mass density of this nanoplate vary continuously in power-law form as

$$\mathscr{P}(z) = (\mathscr{P}_c - \mathscr{P}_m)\left(\frac{z}{h} + \frac{1}{2}\right)^k + \mathscr{P}_m \tag{6.1}$$

where $\mathscr{P}_c$ and $\mathscr{P}_m$ denote the values of the material properties of the ceramic and metal constituents of the FG plate respectively. k (power-law exponent) is a non-negative variable parameter. According to this distribution, the bottom surface ($z = -h/2$) of the FG plate is pure metal, whereas the top surface ($z = h/2$) is pure ceramic and for different values of k one can obtain different volume fractions of material plate. For this study, Young's modulus (E) and mass densities (ρ) are taken into consideration while other properties will remain constant through the thickness of the plate.

6.3 Numerical modeling

First, the classical plate theory in conjunction with nonlocal elasticity theory decides the displacement field of the FG nanoplate. Then the Rayleigh–Ritz method is employed to obtain the generalized eigenvalue problem, which finds the eigenfrequencies of different modes. Let us first consider the displacement field based on classical plate theory as

$$\begin{aligned} u_x(x, y, z) &= -z\frac{\partial w}{\partial x} \\ u_y(x, y, z) &= -z\frac{\partial w}{\partial y} \\ u_z(x, y, z) &= w(x, y) \end{aligned} \tag{6.2}$$

where u_x, u_y, and u_z are the displacement components along x, y, and z coordinate directions respectively and w is the transverse deflection of a point on the mid-plane (x–y plane). Transverse shear deformation is neglected in the case of the Kirchoff assumption, which means the deformation occurred due to bending and in-plane stretching. The nonzero linear strains associated with the displacement field can be expressed as

$$\begin{Bmatrix} \varepsilon_{xx} \\ \varepsilon_{yy} \\ \gamma_{xy} \end{Bmatrix} = \begin{Bmatrix} \dfrac{\partial u_x}{\partial x} \\ \dfrac{\partial u_y}{\partial y} \\ \dfrac{\partial u_x}{\partial y} + \dfrac{\partial u_y}{\partial x} \end{Bmatrix} = \begin{Bmatrix} -z\dfrac{\partial^2 w}{\partial x^2} \\ -z\dfrac{\partial^2 w}{\partial y^2} \\ -2z\dfrac{\partial^2 w}{\partial x \partial y} \end{Bmatrix} \tag{6.3}$$

where ε_{xx} and ε_{yy} are the normal strains in the x- and y-directions respectively and γ_{xy} is the shear strain. Assuming the material constituents of the FG plate obey the generalized Hooke's law, local constitutive relationships can be given as

$$\sigma = \begin{Bmatrix} \sigma_{xx} \\ \sigma_{yy} \\ \tau_{xy} \end{Bmatrix} = \begin{pmatrix} Q_{11} & Q_{12} & 0 \\ Q_{21} & Q_{22} & 0 \\ 0 & 0 & Q_{66} \end{pmatrix} \begin{Bmatrix} \varepsilon_{xx} \\ \varepsilon_{yy} \\ \gamma_{xy} \end{Bmatrix} \tag{6.4}$$

where σ_{xx}, σ_{yy} are the normal stresses and τ_{xy} is the shear stress and the reduced stiffness components, Q_{ij} are given by

$$Q_{11} = Q_{22} = \frac{E(z)}{1 - \nu^2}, \quad Q_{12} = Q_{21} = \frac{\nu E(z)}{1 - \nu^2}, \quad Q_{66} = \frac{E(z)}{2(1 + \nu)}$$

Here, E and ν are Young's modulus and Poisson's ratio of the material constituents respectively. The constitutive relations in terms of nonlocal elasticity theory can be expressed in differential form as

$$(1 - \xi^2 \nabla^2)\sigma^{nl} = \sigma \tag{6.5}$$

with σ and σ^{nl} are local (equation (6.4)) and nonlocal stress tensors, $\nabla^2 = \frac{\partial^2}{\partial x^2} + \frac{\partial^2}{\partial y^2}$ is the Laplacian operator and $\xi = e_0 l$ is nonlocal parameter, which must be chosen suitably to ensure validity of the nonlocal models. The assumed values of this parameter can be found in [5]. Now the stress resultants and bending moments of the nanoplate may be written as

$$\begin{aligned} N_{xx} &= \int_{-h/2}^{h/2} \sigma_{xx}^{nl} dz, \quad N_{yy} = \int_{-h/2}^{h/2} \sigma_{yy}^{nl} dz, \quad N_{xy} = \int_{-h/2}^{h/2} \tau_{xy}^{nl} dz \\ M_{xx} &= \int_{-h/2}^{h/2} z\sigma_{xx}^{nl} dz, \quad M_{yy} = \int_{-h/2}^{h/2} z\sigma_{yy}^{nl} dz, \quad M_{xy} = \int_{-h/2}^{h/2} z\tau_{xy}^{nl} dz \end{aligned} \tag{6.6}$$

Using equations (6.5) and (6.6), the strain and kinetic energies of rectangular nanoplate can be found as

$$U = \frac{1}{2} \int_\Omega \int_{-h/2}^{h/2} \left(\sigma_{xx}^{nl} \varepsilon_{xx} + \sigma_{yy}^{nl} \varepsilon_{yy} + \tau_{xy}^{nl} \gamma_{xy} \right) dz \, d\Omega \tag{6.7}$$

$$T = \frac{1}{2} \int_\Omega \int_{-h/2}^{h/2} \rho(z) \left\{ \left(\frac{\partial u_x}{\partial t} \right)^2 + \left(\frac{\partial u_y}{\partial t} \right)^2 + \left(\frac{\partial u_z}{\partial t} \right)^2 \right\} dz \, d\Omega \tag{6.8}$$

where $d\Omega$ denotes the infinitesimal area element of the FG nanoplate. Referring to [4] and the displacement component as $w(x, y; t) = W(x, y)\exp(i\omega t)$, one can obtain the maximum strain and kinetic energies of FG nanoplate as

$$S = \frac{D_c}{2} \int_0^a \int_0^b D_r \left[W_{xx}^2 + W_{yy}^2 + 2\nu W_{xx} W_{yy} + 2(1 - \nu) W_{xy}^2 \right] dy \, dx \tag{6.9}$$

$$T = \frac{\rho_c h \omega^2}{2} \int_0^a \int_0^b \rho_r \left[W^2 + \xi (W_x^2 + W_y^2) \right] dy\, dx \tag{6.10}$$

where the flexural rigidity of ceramic constituent $D_c = \frac{E_c h^3}{12(1-\nu^2)}$, $\rho_r = 12 \left[\frac{1}{k+1} \left(1 - \frac{1}{\rho_r} \right) + \frac{1}{\rho_r} \right]$ and $D_r = \left[\left\{ \frac{1}{k+3} - \frac{1}{k+2} + \frac{1}{4(k+1)} \right\} \left(1 - \frac{1}{E_r} \right) + \frac{1}{E_r} \right]$. In equations (6.9) and (6.10), one can easily introduce the nondimensional parameters $0 \leqslant X(=x/a) \leqslant 1$ and $0 \leqslant Y(=y/b) \leqslant 1$ and obtain the energy expressions as follows

$$S = \frac{D_c ab}{2a^4} \int_0^1 \int_0^1 D_r \left[W_{xx}^2 + \mu^4 W_{yy}^2 + 2\nu\mu^2 W_{xx} W_{yy} + 2(1-\nu)\mu^2 W_{xy}^2 \right] dY\, dX \tag{6.11}$$

$$T = \frac{\rho_c abh\omega^2}{2} \int_0^1 \int_0^1 \rho_r \left[W^2 + \bar{\xi} (W_x^2 + \mu^2 W_y^2) \right] dY\, dX \tag{6.12}$$

with $\mu = a/b$ (aspect ratio of rectangular nanoplate) and the transformed nonlocal scaling parameter $\bar{\xi} = \left(\frac{\xi}{a} \right)^2$. A change of order of integration will not affect the solution, since we are dealing with rectangular nanoplates. For other geometries, one needs to carefully consider the order and limits of integration here in equations (6.11) and (6.12).

Now as the Rayleigh–Ritz method is a method of successive approximation, one may assume the amplitudes of vibration can be expressed as a linear combination of simple algebraic polynomials as $W(X, Y) = \sum_{i=1}^{n} c_i \phi_i(X, Y)$ with c_i as unknown coefficients to be determined, that decide the mode shapes and ϕ_i are the admissible functions corresponding to the amplitudes. These admissible functions must satisfy essential boundary conditions, which can be represented as $\phi_i(X, Y) = f(X, Y)\psi(X, Y)$; $i = 1, 2, \cdots, n$. The polynomials in this series are considered as simple algebraic polynomials generated from Pascal's triangle. In addition, the function $f(X, Y) = X^p Y^q (1-X)^r (1-Y)^s$ is suitably chosen depending on the geometry of the nanoplate, in which the exponents control the different classical edge supports. The parameter p takes the values as 0, 1, and 2 according to the side as $X=0$ is free (F), simply supported (S) and clamped (C), similar interpretations can be assumed for q, r, and s corresponding to the sides $Y=0$, $X=1$, and $Y=1$ respectively.

The Rayleigh quotient (ω^2) can be obtained by equating the maximum strain and kinetic energies mentioned in equations (6.11) and (6.12). Then taking partial derivatives of ω^2 in terms of constant coefficients can find the generalized eigenvalue problem of the form

$$(\mathbb{K} - \lambda^2 \mathbb{M})\Delta = 0 \tag{6.13}$$

where $\mathbb{K}$ and $\mathbb{M}$ are referred as stiffness and inertia matrices respectively; Δ is the vector of unknown constant coefficients and the eigenvalues ($\lambda = \omega a^2 \sqrt{\frac{\rho_c h}{D_c}}$) are the natural frequencies of the vibrating nanoplate.

6.4 Results and discussion

In this section, the convergence of natural frequencies FG rectangular (or square) nanoplates is achieved by assuming 21 polynomials in the displacement component. Then the nondimensional frequencies are being evaluated by varying certain parameters such as the nonlocal scaling parameter ($\bar{\xi}$), power-law exponents (k), and aspect ratio (μ) in tables 6.1, 6.2, and 6.3 respectively. Referring to Rong *et al* [5], we have found the values of ξ as 0, 1, 2, and 3 nanometer (nm). Though in our study, the values of $\bar{\xi}$ have been taken as 0, 0.01, 0.04, and 0.09 nm since a is assumed as 10 nm in the present modeling. In these computations, the nanoplate is supported only with a few classical edge supports. The validation of fundamental frequencies is carried out in these tabulation with Rong *et al* [5] and our results are

Table 6.1. Effect of $\bar{\xi}$ (in nm) on natural frequencies of isotropic square nanoplates with $k = 0$.

ES	$\bar{\xi}$	λ_1	λ_2	λ_3	λ_4	λ_5	λ_6
CCCC	0	35.9888	73.3989	73.3989	108.2653	131.8982	132.424
	0.01	32.3022	58.4763	58.4763	78.4868	90.7410	91.6222
	0.04	25.6207	40.2843	40.2843	50.2908	56.3873	57.5132
	0.09	20.1100	29.3863	29.3863	35.6976	39.6306	40.7347
SSSS	0	19.7392	49.3490	49.3490	79.4007	100.1729	100.187
	Rong *et al* [5]	19.739 21	—	—	—	—	—
	Singh and Azam [6]	19.736 01	—	—	—	—	—
	0.01	18.0390	40.3811	40.3811	59.3274	70.9515	70.9630
	Rong *et al* [5]	18.038 97	—	—	—	—	—
	Singh and Azam [6]	18.035 92	—	—	—	—	—
	0.04	14.7556	28.6163	28.6163	38.9071	44.9172	44.9252
	Rong *et al* [5]	14.755 56	—	—	—	—	—
	Singh and Azam [6]	14.7442	—	—	—	—	—
	0.09	11.8462	21.1556	21.1556	27.8628	31.7713	31.7771
SCSC	0	28.9509	54.7439	69.3271	94.7034	103.7178	129.312
	Rong *et al* [5]	28.950 86	—	—	—	—	—
	Singh and Azam [6]	28.930 81	—	—	—	—	—
	0.01	26.2157	44.5342	55.5345	69.6954	73.1695	89.5321
	Rong *et al* [5]	26.215 65	—	—	—	—	—
	Singh and Azam [6]	26.182 57	—	—	—	—	—
	0.04	21.1091	31.3615	38.4788	45.1741	46.2033	56.0561
	Rong *et al* [5]	21.109 14	—	—	—	—	—
	Singh and Azam [6]	21.0988	—	—	—	—	—
	0.09	16.7516	23.1151	28.1454	32.2057	32.6522	39.1378
CCCF	0	23.9605	40.0195	63.2914	76.7587	80.6895	117.8576
	0.01	22.2314	33.5003	51.6436	55.9054	61.1678	79.1882
	0.04	18.5455	24.2079	36.1886	36.4638	40.2797	48.7022
	0.09	14.9318	18.2187	26.3130	26.4476	28.9319	34.2519

Table 6.2. Effect of power-law exponents (k) on natural frequencies of square FG nanoplates with $\bar{\xi} = 0.04$ nm.

ES	k	λ_1	λ_2	λ_3	λ_4	λ_5	λ_6
CCCC	0	25.6207	40.2843	40.2843	50.2908	56.3873	57.5132
	0.1	24.7050	38.8446	38.8446	48.4934	54.3721	55.4577
	1	21.3177	33.5186	33.5186	41.8444	46.9171	47.8539
	2	20.3778	32.0408	32.0408	39.9996	44.8486	45.7441
CCCF	0	18.5455	24.2079	36.1886	36.4638	40.2797	48.7022
	0.1	17.8827	23.3427	34.8953	35.1607	38.8402	46.9616
	1	15.4308	20.1422	30.1107	30.3398	33.5148	40.5226
	2	14.7505	19.2541	28.7832	29.0021	32.0372	38.7361
SSSS	0	14.7556	28.6163	28.6163	38.9071	44.9172	44.9252
	0.1	14.2282	27.5936	27.5936	37.5166	43.3119	43.3196
	1	12.2774	23.8102	23.8102	32.3726	37.3734	37.3800
	2	11.7361	22.7604	22.7604	30.9454	35.7256	35.7320
SCSC	0	21.1091	31.3615	38.4788	45.1741	46.2033	56.0561
	0.1	20.3547	30.2407	37.1037	43.5596	44.5521	54.0527
	1	17.5639	26.0944	32.0163	37.5871	38.4435	46.6415
	2	16.7895	24.9439	30.6048	35.9300	36.7486	44.5852

Table 6.3. Effect of aspect ratio (μ) on natural frequencies of rectangular FG nanoplates with $\bar{\xi} = 0.04$ and $k = 1$.

ES	$\bar{\xi}$	λ_1	λ_2	λ_3	λ_4	λ_5	λ_6
CCCC	0.2	15.3090	15.6000	16.1343	16.9575	18.6480	20.9321
	0.5	16.1190	19.0252	23.7631	29.6806	30.8790	32.6367
	1.0	21.3177	33.5186	33.5186	41.8444	46.9171	47.8539
	2.0	43.9402	48.4378	56.9373	67.9558	71.2429	74.1405
CCCF	0.2	15.1468	15.3387	15.6530	16.2494	17.1713	24.1783
	0.5	15.1773	16.2403	18.7759	23.2740	29.7817	30.0133
	1.0	15.4308	20.1422	30.1107	30.3398	33.5148	40.5226
	2.0	17.3773	30.9457	34.1530	43.1158	45.2953	54.8335
SSSS	0.2	7.1909	7.8893	9.0538	10.6352	16.9427	20.5958
	0.5	8.3996	12.2775	17.8288	21.3283	23.9423	24.4289
	1.0	12.2774	23.8102	23.8102	32.3726	37.3734	37.3800
	2.0	23.8098	32.2172	43.4838	50.2739	55.3591	56.3889
SCSC	0.2	7.2438	8.0826	9.3998	11.1103	14.1598	17.3687
	0.5	9.2636	14.4794	20.8957	21.6093	24.8229	27.8315
	1.0	17.5639	26.0944	32.0163	37.5871	38.4435	46.6415
	2.0	43.0725	45.2350	51.3916	61.1596	70.9327	72.9844

found to be in excellent agreement. It is evident that eigenfrequencies follow a descending pattern with increase in $\bar{\xi}$ and k, whereas follow an ascending pattern with increase in aspect ratios.

On the other hand, the parameters assumed in finding first six eigenvectors for a square FG nanoplate are $k = 2$ and $\bar{\xi} = 0.04$, which indeed refer to the first six mode shapes as depicted in figures 6.1 and 6.2. One can easily report the natural frequencies for other sets of boundary supports by satisfying criteria of admissible functions in terms of indices p, q, r, and s and hence the mode shapes too.

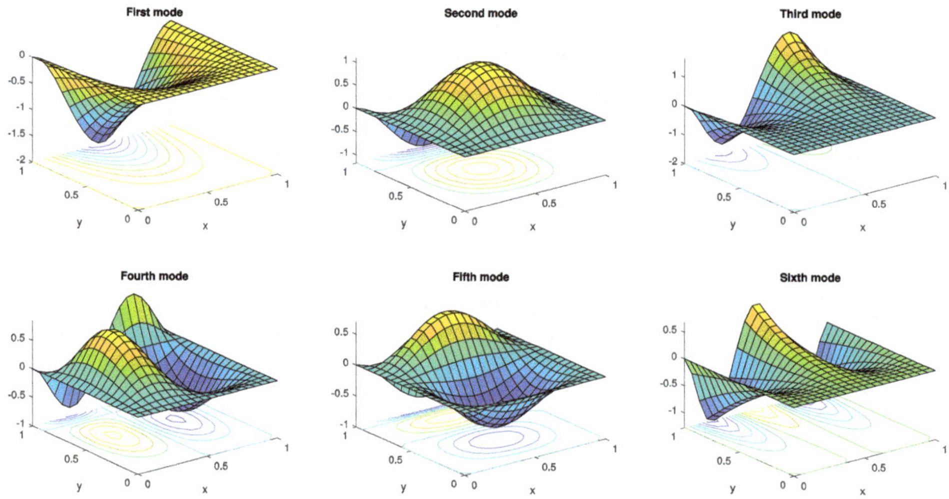

Figure 6.1. First six mode shapes of CCCF square FG nanoplate with $k = 1$ and $\bar{\xi} = 0.04$.

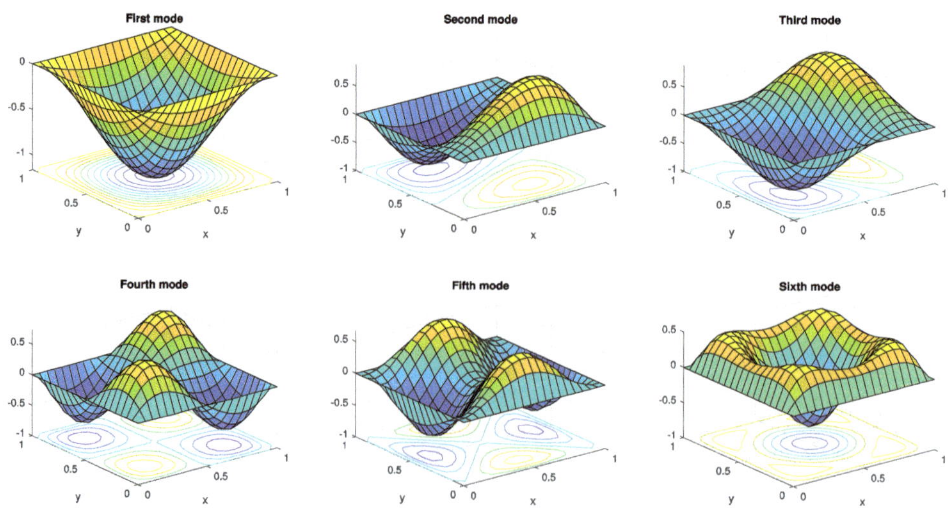

Figure 6.2. First six mode shapes of SSSS square FG nanoplate with $k = 2$ and $\bar{\xi} = 0.04$.

6.5 Concluding remarks

In this investigation, transverse vibration of rectangular FG nanoplate subject to classical boundary conditions is studied based on nonlocal elasticity theory in conjunction with classical plate theory. The Rayleigh–Ritz method (a method of successive approximation) is implemented here to obtain the generalized eigenvalue problem, which helps us finding the natural frequencies along with mode shapes. The benchmark results obtained in this study are summarized below.

- The convergence of eigenfrequencies can be obtained with an increase in the number of polynomials in the displacement component.
- From the validation given in table 6.1, the results found in the given formulation are in excellent agreement with the available literature.
- The natural frequencies are increasing with increase in aspect ratio, whereas they follow a decreasing pattern with increase in power-law indices (k) and nonlocal scaling parameters ($\bar{\xi}$).
- The shear deformation is not included in classical plate theory, but one can easily extend this study of FG nanoplates based on different existing shear deformation theories.

References

[1] Pradhan S C and Phadikar J K 2009 Nonlocal elasticity theory for vibration of nanoplates *J. Sound Vib.* **325** 206–23

[2] Phadikar J K and Pradhan S C 2010 Variational formulation and finite element analysis for nonlocal elastic nanobeams and nanoplates *Comput. Mater. Sci.* **49** 492–9

[3] Guo D, Xie G and Luo J 2014 Mechanical properties of nanoparticles: basics and applications *J. Phys. D: Appl. Phys.* **47** 013001 (25 pages)

[4] Anjomshoa A 2013 Application of Ritz functions in buckling analysis of embedded orthotropic circular and elliptical micro/nano-plates based on nonlocal elasticity theory *Meccanica* **48** 1337–53

[5] Rong D, Fan J, Lim C W, Xu X and Zhou Z 2018 A new analytical approach for free vibration, buckling and forced vibration of rectangular nanoplates based on nonlocal elasticity theory *Int. J. Struct. Stab. Dyn.* **18** 1850055 (27 pages)

[6] Singh P P and Azam M S 2020 Free vibration and buckling analysis of elastically supported transversely inhomogeneous functionally graded nanoplate in thermal environment using Rayleigh-Ritz method *J. Vib. Control.* **27** 1–13

IOP Publishing

Advances in Modeling and Analysis of Functionally Graded Micro- and Nanostructures

Subrat Kumar Jena, S Pradyumna and S Chakraverty

Chapter 7

Static or bending analysis of functionally graded micro- or nanoshells

Shahriar Dastjerdi and Mehran Kadkhodayan

This chapter investigates the nonlinear static bending analysis of functionally graded micro- and nanoshells. The effect of structural porosity defects is also studied. An attempt is made to include the geometry of the analyzed shell in several standard geometries, such as spherical, cylindrical, and conical. Also, moderately thick shell structures and large static deformations are considered. The first-order shear deformation theory (FSDT) has been assumed to obtain the strain components. Also, since the micro- and nanoscale analysis should be done, the modified couple stress theory for the microscale and the nonlocal elasticity theory have been used for the nanoscale analysis. The shell material being studied is functionally graded FGM; hence, the functional changes of material properties along the thickness are considered regarding the power-law index rule. The static governing equations and the mathematical definition of the boundary conditions are derived using Hamilton's energy principle. The static nonlinear governing equations are solved using the semi-analytical polynomial method (SAPM) solution method, and the numerical results are obtained. The obtained numerical results have been evaluated with the results of ABAQUS (a powerful software suite for finite element analysis (FEA) used in engineering simulations). The influence of various factors, including the geometrical shape of the analyzed shell, environmental and structural factors, FGM material, boundary conditions, the effects of small-scale analysis, and several other factors of the results, are drawn and interpreted in several figures. This chapter is suitable as a reference for researchers working on the static analysis of FGM shells at the micro- and nanoscale.

7.1 Introduction

Recently, functionally graded materials (FGM) have received much attention from researchers due to their unique properties [1–3]. The mechanical properties of a

doi:10.1088/978-0-7503-6024-1ch7

structure made of FGM materials, such as Young's modulus, Poisson's ratio, thermal expansion coefficient, etc, will change in one or more directions of the structure, which will cause the material to become inhomogeneous [4–6]. Due to their high strength and geometric curvature, shell-shaped structures can have exceptional and interesting applications if made of FGM materials. Coverings that must be highly resistant to corrosion and wear and electrically insulating connections can be made from FGM materials. In general, more thermal applications are being considered due to the change in the physical properties of FGM structures from ceramic to metal. We can also expect applications such as making sensors and electronic components, often on small micro- and nanoscales. Therefore, the mechanical simulation of FGM shell structures on a small scale is a challenge for researchers that should be considered.

The free vibration analysis of FGM cylindrical shells surrounded by a Pasternak elastic foundation in a thermal environment has been studied by Baghlani *et al* [7]. Also, Kim [8] studied the effect of a partially elastic foundation in the analysis of free vibrations of FGM cylindrical shells by considering the first-order shear deformation theory (FSDT). Bagheri *et al* [9] considered the free vibration of two types of cylindrical and spherical FGM shells. They investigated the effects of the power-law index in their research. A cylindrical shell with an FGM core rotating around its central axis has been investigated by Karroubi and Irani–Rahaghi [10]. Two piezoelectric layers surround the analyzed structure. Other researchers have also considered various issues, including the effect of different boundary conditions, the finite element method in analysis, and variable thickness for FGM cylindrical structures with the vibration approach [11–13]. Shahmohammadi *et al* [14] have investigated the stability of laminated composites and FGM sandwich shells with a new isogeometric method. Considering von Karman's theory, the nonlinear behavior of cylindrical and spherical FGM panels has been modeled by Amieur *et al* [15]. Esmaeili *et al* [16] have researched the nonlinear vibrations caused by the rapid heating of cylindrical FGM shells. They have used the Newmark time marching approach and the Picard method to solve the obtained equations. Nonlinear analyses and the thermal post-buckling study of FGM shell structures have also attracted the attention of researchers [17, 18]. Until now, most of the mentioned research has considered cylindrical or spherical shell structures. However, Sofiyev [19] has studied the stability and vibrations of FGM conical shells. He has considered the Galerkin solution method to solve the governing equations. Dastjerdi and Akgöz [20] have performed an exact three-dimensional analysis for static bending and free vibrations of annular and circular FGM macro and nanosheets in a thermal environment. In this analysis, when the strains along the thickness are present, they concluded that unacceptable results would be obtained if they were omitted in the thermal analysis. Also, they have presented several studies on the mechanical behavior analysis of FGM shell structures, including viscoelastic FGM gyroscopes, functional cross-section FGM shell structures, and thick FGM plates [21–23].

A static analysis of the bending and buckling of an FGM microcylinder with variable thickness was done by Ghareghani *et al* [24]. They have used modified couple stress theory (MCST) to simulate analysis at the microscale.

Sladek *et al* [25] have also used the couple stress theory to analyze the bending behavior of micro-sized sheets. They have done an in-depth study on the effect of the small-scale parameter on the results. The study of flexoelectric and piezoelectric effects of FGM conical nanoshells has been considered by Khorshidi *et al* [26]. They used Eringen's nonlocal elasticity theory and FSDT approach to simulate nanosize effects. Buckling and post-buckling analysis of FGM nanoshells modeled by nonlocal strain gradient theory have been investigated by Sahmani and Fattahi [27]. They concluded that nonlocal small-scale effects on FGM nanoshell instability are more significant than strain gradient effects. Yang *et al* [28] have studied the nonlinear post-buckling behavior of FGM microshells under hydrostatic pressure using the nonlocal strain gradient approach. They have concluded that nonlocal effects will reduce the critical hydrostatic pressure. Shojaeefard *et al* [29] have researched the piezomagnetic analysis of vibrations and buckling of rotating FGM nanoshells. Their analyzed nanoshell rests on a viscoelastic foundation. One of the results they have obtained is that the free vibrations of the piezomagnetic FGM material in the cylindrical nanoshell structure are influenced by several different factors, including angular velocity, length scale parameter, external voltage and amperage, parameters related to viscoelastic foundation and also related FGM power index factor. It is observed that most of the mentioned research studies have a vibration or buckling stability approach. Regarding fully static analysis, Ma *et al* [30] have researched large nonlinear quasi-3D deformations for microplates with variable thickness by applying the nonlocal stress–strain gradient theory.

The imperfection analysis of FGM nanoshells with the instability approach has also been researched [31, 32]. Porosity is one of the things that will cause structural defects and weaken the strength of the structure. Timesli [33] presented an analytical model for the porous behavior of cylindrical FGM shells resting on an elastic matrix. He has chosen the Winkler–Pasternak model to simulate the elastic foundation. Sheykhi *et al* [34] investigated the nonlinear free vibrations of truncated conical nanoshells using modified strain gradient theory. They showed that the degree of hardening of the nanoshell in the modified strain gradient theory is lower than in the classical and the modified couple stress theories. The Euler–Bernoulli beam model for a nanoscale beam with changing properties in 3D using couple stress theory has been studied by Hadi *et al* [35]. They have used the GDQM solving method to solve the obtained governing equations. Jung and Han [36] have considered the nonlocal elasticity approach for analyzing nanoplates made of sigmoid functionally graded material. They have derived the equilibrium equations using Hamilton's principle. Another study on the dynamic analysis of FGM porous nanoshells was done by Ghandourah *et al* [37]. They have solved the derived dynamic equations using an analytical method based on the Galerkin solution method. Much research has been conducted on the analysis of FGM nanoplates, among which one can mention the vibration analysis of FGM annular nanosheets, which is modeled by the first-order shear deformation theory of the plates [38]. One of the related interesting studies is the research done by Monge and Mantari [39]. They have presented a three-dimensional analysis of the static behavior of FGM shells. Although this research has been done on a macro scale, since a doubly-curved FGM shell structure has been considered, it is complete in terms of the geometrical shape of the analyzed shell. So, the results of various shell structures can be obtained from this simulation. Also, they have paid attention to the differential quadrature method (DQM) solution method to solve the extracted equations. Dehsaraji *et al* [40] have investigated the effects of thickness stretching in analyzing free vibrations of piezoelectric FGM micro- and nanoshells. They used the modified couple stress theory to simulate the size-dependent behavior of the analyzed structure. Hosseini–Hashemi *et al* [41] have presented the free vibration analysis of micro- and nanospherical shells based on the modified couple stress theory. In the practical justification of selecting the spherical nanoshell, they gave the example of fullerene carbon structure C_{60}. Also, they and other researchers have used the plate theory of FSDT to analyze micro- and nanoFGM shells [42].

According to the previous research, the analysis of the mechanical behavior of FGM micro-and nanoshells is of particular importance. Therefore, in this chapter, an attempt has been made to comprehensively analyze the static analysis of micro- and nanoFGM shells. This chapter considers two conical and spherical FGM shells embedded on the Winkler–Pasternak type elastic foundation. The material's functionally graded property is considered a power-law index along the thickness from metal to ceramic. Also, the FSDT theory, along with two modified couple stress and Eringen's nonlocal elasticity theories, have been used to simulate the behavior of the micro- and nanoscale structure. In order to obtain large deformations, von Karman's assumptions have been considered. The static governing equations and description of the boundary conditions are derived by applying the energy method. The obtained equations have been solved with a semi-analytical method based on polynomials called SAPM. Different types of boundary conditions and loads have been taken into consideration. In the following, the effect of essential parameters on the results, especially the parameter related to the small-scale effects, has been investigated. The discussed problem in this chapter can be a good guide for researchers who study the static analysis of micro- and nanoFGM shells.

7.2 The geometry of the problem

The geometry of the analyzed FGM structure in this chapter is considered as shells. The shell structures cover a wide range of structures used in industry. Shells can withstand more loads than sheets due to their initial curvature before loading. In other words, regardless of the material properties of the structure, shells can withstand membrane stresses due to their specific geometric shape. For example, in a spherical geometry under uniform internal pressure, if the radius of the sphere is 20 times its thickness, the ratio of membrane stresses to bending stresses is about 40 times. In other words, there is no need to analyze by considering the material of the structure and the bending stresses. Therefore, the ratio parameter of the radius R to the thickness h is critical and determines how accurate the results are if we only

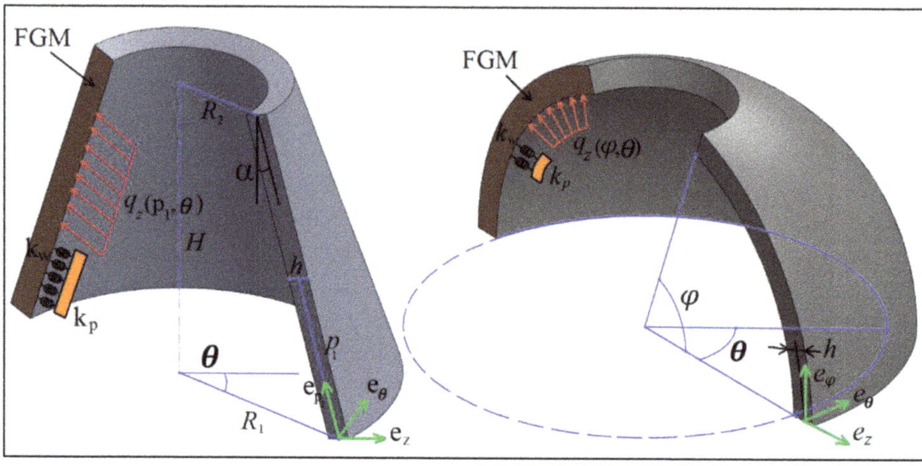

Figure 7.1. Schematic view of conical and spherical micro- and nanoFGM shells.

consider the membrane stresses in the structure. In this chapter, two structures that are more useful in industry and reality, such as spherical and conical structures, have been considered. The cylindrical structure can be easily simulated by considering a conical geometry that will be discussed. The geometric schematic shape of the analyzed structures can be seen in figure 7.1. The two analyzed structures are subjected to the transverse nonuniform load. The suitable coordinate system for each structure is introduced in figure 7.1, respectively spherical and conical coordinate systems. According to the nature of the FGM, it can be seen that the properties of the material change from metal to ceramic along the thickness. The specifications of the geometric dimensions of the two shell structures can be seen in figure 7.1 ($0 \leqslant p_1 \leqslant L$).

7.3 Functionally graded material

The material of the analyzed structure in this chapter is FGM. FGM is a material whose properties change in different directions or a specific direction as a function of a value on one surface to a different value on another surface. Therefore, the material properties in FGM structures are inhomogeneous. For example, if the material properties along the thickness (h) of an arbitrary shell structure, as shown in figure 7.1, change from metal properties at $z = -\frac{h}{2}$ to ceramic material properties at $= +\frac{h}{2}$, a function like equation (7.1) can be introduced for Young's modulus.

$$E(z) = (E_c - E_m)\left(\frac{z}{h} + \frac{1}{2}\right)^g + E_m\left(-\frac{h}{2} \leqslant z \leqslant \frac{h}{2}\right) \tag{7.1}$$

Other properties will be changed as equation (7.1). For example, the Poisson ratio $\nu(z)$ and the thermal expansion coefficient $\alpha(z)$ will be formulated as follows:

$$\nu(z) = (\nu_c - \nu_m)\left(\frac{z}{h} + \frac{1}{2}\right)^g + \nu_m \left(-\frac{h}{2} \leqslant z \leqslant \frac{h}{2}\right) \tag{7.2}$$

$$\alpha(z) = (\alpha_c - \alpha_m)\left(\frac{z}{h} + \frac{1}{2}\right)^g + \alpha_m \left(-\frac{h}{2} \leqslant z \leqslant \frac{h}{2}\right) \tag{7.3}$$

In equation (7.1), values of E_c and E_m are Young's modulus of fully ceramic and metal surfaces at $z = -\frac{h}{2}$ and $z = +\frac{h}{2}$. In the interval $-\frac{h}{2} \leqslant z \leqslant \frac{h}{2}$, if we want to obtain Young's modulus, equation (7.1) should be used. The g parameter that appeared as a power actually shows the rate of change of properties from ceramic to metal here (power-law index). For example, if $g = 0$, the entire structure has ceramic properties, and if g has a tremendous value (∞), the properties of the structure tend to be metal properties. To better understand the effect of parameter g in FGM materials, the changes of Young's modulus along the thickness for a material with the following characteristics are drawn in figure 7.2. In the numerical results section of this chapter, the effect of parameter g on the results will be discussed in detail. The critical point, according to the FGM, is that FGMs are completely nonhomogeneous, and their mechanical analyses are not as simple as the isotropic or even orthotropic structures. Consequently, unique solution methods (usually numerical methods) should be used to solve the governing equations of FGM structures.

$$E_c = 385 \text{ GPa};\ E_m = 210 \text{ GPa};\ h = 0.1 \text{ m};\ 0 \leqslant g \leqslant 10 \tag{7.4}$$

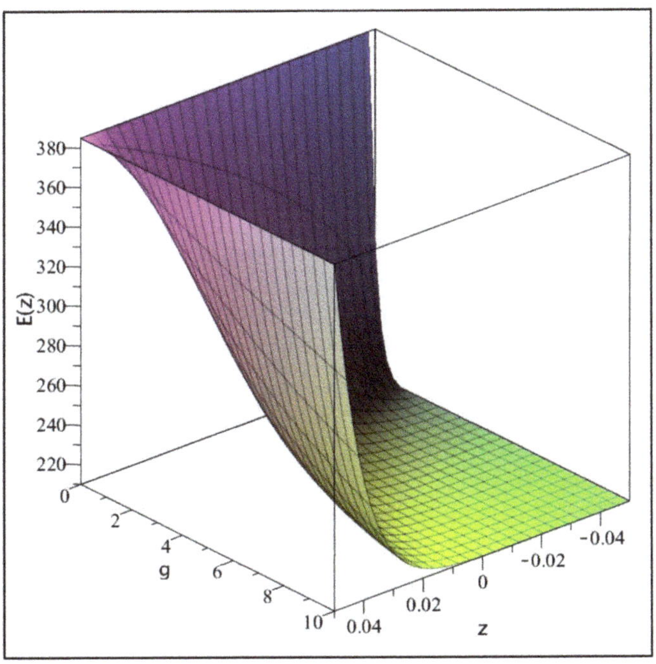

Figure 7.2. The Young's modulus variations of a FGM structure versus g and z.

7.4 Micro- and nanosize structures

In this chapter, the discussed FGM shell structure is in the micro- 10^{-6} m and nano- 10^{-9} m size range. Mechanical analysis on a macroscopic scale structure can be done according to known theories in this regard. However, on small scales, especially on the nanoscale, other well-known theories that locally consider stress and strain at a point under the influence of local factors at the same position will be invalid. The reason for this topic is the presence of other forces whose shallow impact can be ignored on a macroscopic scale. However, if, for example, we analyze a structure on the scale of a few nanometers mechanically, we must recognize the effects of stretching, bending, and twisting caused by atomic forces. The best analysis in small scales (which, of course, is also accurate for macroscopic scales) is the experimental results. However, as it is evident, the costs of conducting experimental tests, especially at the atomic and nanometer scales, are very high. Also, the conditions for conducting such tests are another discussion, which will lead to inaccurate results if exact laboratory conditions are not provided in atomic scales.

Therefore, researchers have proposed other methods of analyzing small-scale structures, including molecular dynamics, structural mechanics, and small-scale continuum mechanics. The mentioned methods have been listed for obtaining more accurate results and percentage of difficulty, respectively. For example, molecular dynamics provides very accurate results. However, analyzing the behavior of each atom separately, considering all the atoms in a structure, and analyzing the effects of all of them for a loading condition can greatly slow down the time to achieve results. Therefore, potent computer processors should be used in this regard. As a result, like the laboratory methods, the cost of simulation is moderately high regarding molecular dynamics. In the structural mechanics method, which connects and simulates potential atomic energies with well-known mechanical structures such as trusses, springs, dampers, frames, beams, etc, suitable results can be obtained. However, this method is more helpful in analyzing a specific problem, and it will slow down the analysis process to simulate a wide range of structures (e.g., in this chapter, where several shell structures must be analyzed). The third method, which is the small-scale continuum mechanics, is a method with simpler equations than the other previously mentioned methods. Of course, the provided results according to this method are less accurate than the methods of molecular dynamics and structural mechanics. However, it provides a very good perspective and understanding of how a structure behaves on a small scale. Among the small-scale continuum mechanics methods, we can mention nonlocal elasticity theory, modified couple stress theory, and strain gradient theory as well. In this chapter, the nonlocal elasticity theory is discussed in detail, and the nonlinear bending analysis for an FGM shell structure has been done according to this theory. Also, the modified couple stress theory for the analysis of small-scale structures is briefly mentioned.

7.5 The nonlocal elasticity theory

In the macroscopic scale, stress and strain have a local definition. In other words, the strain at one point of the structure depends on the stress at the same point. However, in Eringen's nonlocal elasticity, stress and strain have a nonlocal definition. For example, the strain at a specific point is affected by the strain of all points in the geometry domain of the problem. Eringen has presented an integral and differential equation in this regard as follows.

$$\sigma_{ij}(x) = \int_V \lambda(|x - x'|, \eta) C_{ijkl} \varepsilon_{kl}(x') dV(x'), \ \forall \ x \in V \tag{7.5}$$

$$(1 - \mu \nabla^2) \sigma_{ij}^{NL} = \sigma_{ij}^{L} = C\text{:}\ \varepsilon, \ \ \mu = (e_0 a)^2 \tag{7.6}$$

In the above equations, $\sigma_{ij}(x)$, $\lambda(|x - x'|, \eta)$, $|x - x'|$, ε_{ij} and C_{ijkl} are the nonlocal stress, modulus, the line distance between point x and x', strain, and general stiffness matrix, respectively. η can be determined as $(e_0 a/l)$. Also, in equation (7.6), σ_{ij}^{NL} are the nonlocal stresses, and σ_{ij}^{L} the local stresses that have a macroscopic definition. The parameter $e_0 a$ is the nonlocal coefficient. e_0 is a dimensionless parameter that depends on the type of material, and a is a scale-dependent parameter (with a dimension of meter) that depends on many different atomic factors. The nonlocal coefficient is also defined as $\mu = (e_0 a)^2$. According to the definition of nonlocal stresses (especially the differential equation) based on its dependence on local stresses and the nonlocal coefficient, the governing equations of a small-scale mechanical structure can be derived. ∇^2 is the Laplacian operator, which has its definition in each coordinate system. We note that the existence of the ∇^2 operator shows the dependence on the entire domain of the problem geometry. For example, in Laplace's famous equation for heat transfer, $\nabla^2 T = 0$ (where T is temperature) will determine the value of T at any point of the problem's geometry domain. If the Cartesian coordinate system is considered, we will have $\left(\frac{\partial^2}{\partial x^2} + \frac{\partial^2}{\partial y^2}\right) T(x, y) = 0$. In the following, the application of nonlocal elasticity theory will be introduced in the nonlinear bending analysis of FGM shell structures at the micro- or nanoscale. Also, the modified couple stress theory will be mentioned briefly.

7.6 Deriving the governing equations

7.6.1 Displacement field

In this section, the method of deriving the governing equations will be explained in detail for FGM shell structures at the micro- and nanoscale. First, regardless of any geometry of the shell structure, the displacements of an arbitrary point should be introduced. The values of U, V, and W represent the displacement of an arbitrary point A on the structure after deformation. If the plate and shell theories are used, the strains along the thickness are ignored due to the low ratio of thickness to other dimensions of the structure. Classical theory is suitable for the analysis of thin plates. However, in this chapter, the final goal is to present the results with high accuracy.

Therefore, the theory will be used that includes shear forces. Therefore, due to considering the relative advantages, the FSDT has been used in this chapter to formulate the displacement field (functions U, V, and W). In FSDT theory, five functions are assumed to introduce U, V, and W functions. The three functions $u_0(r, s)$, $v_0(r, s)$ and $w_0(r, s)$ are transport displacement functions, and φ_r, φ_s are rotation functions around s and r axes in an arbitrary coordinate system (r, s, z) (z represents the coordinate in the thickness direction). Therefore, the displacement field based on FSDT theory is introduced as the following equations

$$\begin{cases} U(r, s, z) = u_0(r, s) + z \cdot \psi_r(r, s) \\ V(r, s, z) = v_0(r, s) + z \cdot \psi_s(r, s) \\ W(r, s, z) = w_0(r, s) \end{cases} \tag{7.7}$$

In FSDT theory, shear force variations along the thickness are introduced as a linear function. One of the weaknesses of this theory is the presence of shear force in the upper and lower surface of the structure, which should be zero due to the boundary conditions in these two areas. The distribution of shear force along the thickness should be changed in the form of a curve that has values equal to zero in the upper and lower levels. To correct this defect, the shear force correction factor κ_s is applied in the calculations related to shear stresses, which is introduced as $\kappa_s = \left(\frac{5}{6}\right)$ for plates.

7.6.2 Spherical and conical shell structure and strain tensors

Now, to continue the process of obtaining the governing equations, the geometry type of the analyzed structure should be determined. First, the strain tensor and its components will be obtained for the analyzed structure according to the introduced displacement field. Two types of asymmetric spherical and conical shell geometries have been considered (according to figure 7.1). The strain tensor according to both types of structure will be presented as the following equations. The coordinate system is spherical (φ, θ, r), and conical (p_1, z, θ)

Spherical structure:

$$\begin{cases} X = r \sin\varphi \cos\theta \\ Y = r \sin\varphi \sin\theta \\ Z = r \cos\varphi \end{cases} \tag{7.8}$$

$$\begin{cases} \vec{\nabla} = \dfrac{\partial}{\partial r}\hat{e}_r + \dfrac{1}{r}\dfrac{\partial}{\partial \varphi}\hat{e}_\varphi + \dfrac{1}{r\sin\varphi}\dfrac{\partial}{\partial \theta}\hat{e}_\theta \\ \nabla^2 = \left(\dfrac{\partial^2}{\partial r^2} + \dfrac{2}{r}\dfrac{\partial}{\partial r}\right) + \dfrac{1}{r^2\sin\varphi}\dfrac{\partial}{\partial\varphi}\left(\sin\varphi\dfrac{\partial}{\partial\varphi}\right) + \dfrac{1}{r^2\sin^2\varphi}\dfrac{\partial^2}{\partial\theta^2} \end{cases} \tag{7.9}$$

$$\begin{bmatrix} \dfrac{\partial \hat{\boldsymbol{e}}_r}{\partial r} = 0 & \dfrac{\partial \hat{\boldsymbol{e}}_r}{\partial \varphi} = \hat{\boldsymbol{e}}_\varphi & \dfrac{\partial \hat{\boldsymbol{e}}_r}{\partial \theta} = \sin\varphi \hat{\boldsymbol{e}}_\theta \\ \dfrac{\partial \hat{\boldsymbol{e}}_\varphi}{\partial r} = 0 & \dfrac{\partial \hat{\boldsymbol{e}}_\varphi}{\partial \varphi} = -\hat{\boldsymbol{e}}_r & \dfrac{\partial \hat{\boldsymbol{e}}_\varphi}{\partial \theta} = \cos\varphi \hat{\boldsymbol{e}}_\theta \\ \dfrac{\partial \hat{\boldsymbol{e}}_\theta}{\partial r} = 0 & \dfrac{\partial \hat{\boldsymbol{e}}_\theta}{\partial \varphi} = 0 & \dfrac{\partial \hat{\boldsymbol{e}}_\theta}{\partial \theta} = -(\sin\varphi \hat{\boldsymbol{e}}_r + \cos\varphi \hat{\boldsymbol{e}}_\varphi) \end{bmatrix} \tag{7.10}$$

$$\overleftrightarrow{\varepsilon} = \frac{1}{2}\left[\nabla \vec{U} + \nabla\, \vec{U}^T + \nabla\, \vec{U} \cdot \nabla\, \vec{U}^T\right] \tag{7.11}$$

$$\overleftrightarrow{\varepsilon}_{ij} = \begin{bmatrix} \varepsilon_{rr} & \varepsilon_{r\varphi} & \varepsilon_{r\theta} \\ \varepsilon_{\varphi r} & \varepsilon_{\varphi\varphi} & \varepsilon_{\varphi\theta} \\ \varepsilon_{\theta r} & \varepsilon_{\theta\varphi} & \varepsilon_{\theta\theta} \end{bmatrix} \tag{7.12}$$

$$\begin{aligned} \nabla \vec{U} &= \begin{bmatrix} \dfrac{\partial}{\partial r}\hat{e}_r & \dfrac{1}{R}\dfrac{\partial}{\partial \varphi}\hat{e}_\varphi & \dfrac{1}{R\sin\varphi}\dfrac{\partial}{\partial \theta}\hat{e}_\theta \end{bmatrix} \begin{bmatrix} W\hat{e}_r \\ U\hat{e}_\varphi \\ V\hat{e}_\theta \end{bmatrix} \\ &= \begin{bmatrix} \dfrac{\partial W}{\partial r} & \dfrac{\partial U}{\partial r} & \dfrac{\partial V}{\partial r} \\ \dfrac{1}{R}\left(\dfrac{\partial W}{\partial \varphi} - U\right) & \dfrac{1}{R}\left(\dfrac{\partial U}{\partial \varphi} + W\right) & \dfrac{1}{R}\dfrac{\partial V}{\partial \varphi} \\ \dfrac{1}{R}\left(\dfrac{1}{\sin\varphi}\dfrac{\partial W}{\partial \theta} - V\right) & \dfrac{1}{R}\left(\dfrac{1}{\sin\varphi}\dfrac{\partial U}{\partial \theta} - V\cot\varphi\right) & \dfrac{1}{R}\left(W + U\cot\varphi + \dfrac{1}{\sin\varphi}\dfrac{\partial V}{\partial \theta}\right) \end{bmatrix} \end{aligned} \tag{7.13}$$

$$\nabla \vec{U}^T = \begin{bmatrix} \dfrac{\partial W}{\partial r} & \dfrac{1}{R}\left(\dfrac{\partial W}{\partial \varphi} - U\right) & \dfrac{1}{R}\left(\dfrac{1}{\sin\varphi}\dfrac{\partial W}{\partial \theta} - V\right) \\ \dfrac{\partial U}{\partial r} & \dfrac{1}{R}\left(\dfrac{\partial U}{\partial \varphi} + W\right) & \dfrac{1}{R}\left(\dfrac{1}{\sin\varphi}\dfrac{\partial U}{\partial \theta} - V\cot\varphi\right) \\ \dfrac{\partial V}{\partial r} & \dfrac{1}{R}\dfrac{\partial V}{\partial \varphi} & \dfrac{1}{R}\left(W + U\cot\varphi + \dfrac{1}{\sin\varphi}\dfrac{\partial V}{\partial \theta}\right) \end{bmatrix} \tag{7.14}$$

$$\varepsilon_{rr} = \left(\frac{\partial W}{\partial r}\right) + \frac{1}{2}\left(\frac{\partial W}{\partial r}\right)^2 \tag{7.15}$$

$$2\varepsilon_{r\varphi} = 2\varepsilon_{\varphi r} = \frac{\partial U}{\partial r} + \frac{1}{R}\left(\frac{\partial W}{\partial \varphi} - U\right) + \frac{1}{R}\left(\frac{\partial W}{\partial r}\right)\left(\frac{\partial W}{\partial \varphi}\right) \tag{7.16}$$

$$2\varepsilon_{r\theta} = 2\varepsilon_{\theta r} = \frac{\partial V}{\partial r} + \frac{1}{R}\left(\frac{1}{\sin\varphi}\frac{\partial W}{\partial \theta} - V\right) + \frac{1}{R\sin\varphi}\left(\frac{\partial W}{\partial r}\right)\left(\frac{\partial W}{\partial \theta}\right) \tag{7.17}$$

$$\varepsilon_{\varphi\varphi} = \frac{1}{R}\left(\frac{\partial U}{\partial \varphi} + W\right) + \frac{1}{2R^2}\left(\frac{\partial W}{\partial \varphi}\right)^2 \tag{7.18}$$

$$2\varepsilon_{\varphi\theta} = 2\varepsilon_{\theta\varphi} = \frac{1}{R}\left(\frac{\partial V}{\partial \varphi} + \frac{1}{\sin\varphi}\frac{\partial U}{\partial \theta} - V\cot\varphi\right) + \frac{1}{R^2\sin\varphi}\left(\frac{\partial W}{\partial \varphi}\right)\left(\frac{\partial W}{\partial \theta}\right) \tag{7.19}$$

$$\varepsilon_{\theta\theta} = \frac{1}{R}\left(W + U\cot\varphi + \frac{1}{\sin\varphi}\frac{\partial V}{\partial \theta}\right) + \frac{1}{2R^2\sin^2\varphi}\left(\frac{\partial W}{\partial \theta}\right)^2 \tag{7.20}$$

The FSDT displacement field is defined for spherical structure as follows

$$\begin{cases} U(\varphi, \theta, z) = u_0(\varphi, \theta) + z \cdot \psi_\theta(\varphi, \theta) \\ V(\varphi, \theta, z) = v_0(\varphi, \theta) + z \cdot \psi_\varphi(\varphi, \theta) \\ W(\varphi, \theta, z) = w_0(\varphi, \theta) \end{cases} \tag{7.21}$$

Now, by substituting the FSDT displacement field (equation (7.21)) into equations (7.8)–(7.20) strain components for spherical structure will be formulated below. Because of the thin structure ($z \ll R$) we assume $r = R + z$ as R and $\varepsilon_{zz} = \frac{\partial w_0}{\partial(R+z)} \cong 0$

$$\begin{cases} \vec{\nabla} = \frac{\partial}{\partial z}\hat{\boldsymbol{e}}_z + \frac{1}{R}\frac{\partial}{\partial \varphi}\hat{\boldsymbol{e}}_\varphi + \frac{1}{R\sin\varphi}\frac{\partial}{\partial \theta}\hat{\boldsymbol{e}}_\theta \\ \nabla^2 = \left(\frac{\partial^2}{\partial z^2} + \frac{2}{R}\frac{\partial}{\partial z}\right) + \frac{1}{R^2\sin\varphi}\frac{\partial}{\partial \varphi}\left(\sin\varphi\frac{\partial}{\partial \varphi}\right) + \frac{1}{R^2\sin^2\varphi}\frac{\partial^2}{\partial \theta^2} \end{cases} \tag{7.22}$$

$$\left(\overleftrightarrow{\varepsilon}_{ij}\right)_{\text{Spherical}} = \begin{bmatrix} \varepsilon_{zz} = 0 & \varepsilon_{z\varphi} & \varepsilon_{z\theta} \\ \varepsilon_{z\varphi} & \varepsilon_{\varphi\varphi} & \varepsilon_{\varphi\theta} \\ \varepsilon_{z\theta} & \varepsilon_{\theta\varphi} & \varepsilon_{\theta\theta} \end{bmatrix} \tag{7.23}$$

$$\nabla\vec{U} = \begin{bmatrix} \frac{\partial}{\partial z}\hat{e}_z & \frac{1}{R}\frac{\partial}{\partial \varphi}\hat{e}_\varphi & \frac{1}{R\sin\varphi}\frac{\partial}{\partial \theta}\hat{e}_\theta \end{bmatrix} \begin{bmatrix} w_0(\varphi,\theta)\hat{e}_z \\ \left(u_0(\varphi,\theta) + z\cdot\psi_\theta(\varphi,\theta)\right)\hat{e}_\varphi \\ \left(v_0(\varphi,\theta) + z\cdot\psi_\varphi(\varphi,\theta)\right)\hat{e}_\theta \end{bmatrix} =$$

$$\begin{bmatrix} 0 & \psi_\theta & \psi_\varphi \\ \frac{1}{R}\left(\frac{\partial w_0}{\partial \varphi} - u_0 - z\cdot\psi_\theta\right) & \frac{1}{R}\left(\frac{\partial u_0}{\partial \varphi} + z\frac{\partial \psi_\theta}{\partial \varphi} + w_0\right) & \frac{1}{R}\left(\frac{\partial v_0}{\partial \varphi} + z\frac{\partial \psi_\varphi}{\partial \varphi}\right) \\ \frac{1}{R}\left(\frac{1}{\sin\varphi}\frac{\partial w_0}{\partial \theta} - v_0 - z\cdot\psi_\varphi\right) & \frac{1}{R}\left(\frac{1}{\sin\varphi}\left(\frac{\partial u_0}{\partial \theta} + z\frac{\partial \psi_\theta}{\partial \theta}\right) - \left(v_0 + z\cdot\psi_\varphi\right)\cot\varphi\right) & \frac{1}{R}\left(w_0 + \left(u_0 + z\cdot\psi_\theta\right)\cot\varphi + \frac{1}{\sin\varphi}\left(\frac{\partial v_0}{\partial \theta} + z\frac{\partial \psi_\varphi}{\partial \theta}\right)\right) \end{bmatrix} \tag{7.24}$$

$$\varepsilon_{zz} = 0 \tag{7.25}$$

$$2\varepsilon_{z\varphi} = 2\varepsilon_{\varphi z} = \psi_\theta + \frac{1}{R}\left(\frac{\partial w_0}{\partial \varphi} - u_0 - z \cdot \psi_\theta\right) \tag{7.26}$$

$$2\varepsilon_{z\theta} = 2\varepsilon_{\theta z} = \psi_\varphi + \frac{1}{R}\left(\frac{1}{\sin\varphi}\frac{\partial w_0}{\partial \theta} - v_0 - z \cdot \psi_\varphi\right) \tag{7.27}$$

$$\varepsilon_{\varphi\varphi} = \frac{1}{R}\left(\frac{\partial u_0}{\partial \varphi} + z\frac{\partial \psi_\theta}{\partial \varphi} + w_0\right) + \frac{1}{2R^2}\left(\frac{\partial w_0}{\partial \varphi}\right)^2 \tag{7.28}$$

$$\begin{aligned} 2\varepsilon_{\varphi\theta} = 2\varepsilon_{\theta\varphi} = & \frac{1}{R}\left(\frac{\partial v_0}{\partial \varphi} + z\frac{\partial \psi_\varphi}{\partial \varphi} + \frac{1}{\sin\varphi}\left(\frac{\partial u_0}{\partial \theta} + z\frac{\partial \psi_\theta}{\partial \theta}\right) - (v_0 + z \cdot \psi_\varphi)\cot\varphi\right) \\ & + \frac{1}{R^2 \sin\varphi}\left(\frac{\partial W}{\partial \varphi}\right)\left(\frac{\partial W}{\partial \theta}\right) \end{aligned} \tag{7.29}$$

$$\begin{aligned} \varepsilon_{\theta\theta} = & \frac{1}{R}\left(w_0 + (u_0 + z \cdot \psi_\theta)\cot\varphi + \frac{1}{\sin\varphi}\left(\frac{\partial v_0}{\partial \theta} + z\frac{\partial \psi_\varphi}{\partial \theta}\right)\right) \\ & + \frac{1}{2R^2 \sin^2\varphi}\left(\frac{\partial w_0}{\partial \theta}\right)^2 \end{aligned} \tag{7.30}$$

Also, the conical strain tensor in conical coordinate system (p_1, z, θ) can be expressed according to FSDT

$$\begin{cases} U(p_1, z, \theta) = u_0(p_1, \theta) + z \cdot \psi_\theta(p_1, \theta) \\ V(p_1, z, \theta) = v_0(p_1, \theta) + z \cdot \psi_{p_1}(p_1, \theta) \\ W(p_1, z, \theta) = w_0(p_1, \theta) \end{cases} \tag{7.31}$$

$$\left(\overleftrightarrow{\varepsilon}_{ij}\right)_{\text{Conical}} = \begin{bmatrix} \varepsilon_{p_1p_1} & \varepsilon_{zp_1} & \varepsilon_{p_1\theta} \\ \varepsilon_{zp_1} & \varepsilon_{zz} & \varepsilon_{z\theta} \\ \varepsilon_{p_1\theta} & \varepsilon_{z\theta} & \varepsilon_{\theta\theta} \end{bmatrix} \tag{7.32}$$

$$\begin{bmatrix} \dfrac{\partial \hat{e}_{p_1}}{\partial p_1} = 0 & \dfrac{\partial \hat{e}_{p_1}}{\partial z} = 0 & \dfrac{\partial \hat{e}_{p_1}}{\partial \theta} = -(\sin\alpha)\hat{e}_\theta \\ \dfrac{\partial \hat{e}_z}{\partial p_1} = 0 & \dfrac{\partial \hat{e}_z}{\partial z} = 0 & \dfrac{\partial \hat{e}_z}{\partial \theta} = (\cos\alpha)\hat{e}_\theta \\ \dfrac{\partial \hat{e}_\theta}{\partial p_1} = 0 & \dfrac{\partial \hat{e}_\theta}{\partial z} = 0 & \dfrac{\partial \hat{e}_\theta}{\partial \theta} = (\sin\alpha)\hat{e}_{p_1} - (\cos\alpha)\hat{e}_z \end{bmatrix} \tag{7.33}$$

$$\vec{\nabla} = \left[\hat{e}_{p_1}\left(\frac{\partial}{\partial p_1}\right) \ \hat{e}_z\left(\frac{\partial}{\partial z}\right) \ \hat{e}_\theta \frac{1}{(R_1 + z\cos\alpha - p_1 \sin\alpha)}\left(\frac{\partial}{\partial \theta}\right) \right] \tag{7.34}$$

$$\nabla^2 = \frac{\partial^2}{\partial p_1^2} + \frac{-\sin\alpha}{(R_1 + z\cos\alpha - p_1\sin\alpha)}\frac{\partial}{\partial p_1} + \frac{\partial^2}{\partial z^2} + \frac{\cos\alpha}{(R_1 + z\cos\alpha - p_1\sin\alpha)}\frac{\partial}{\partial z} + \frac{1}{(R_1 + z\cos\alpha - p_1\sin\alpha)^2}\frac{\partial^2}{\partial \theta^2} \tag{7.35}$$

$$\begin{aligned} (\varepsilon_{p_1p_1})^{\text{linear}} &= \frac{\partial u_0}{\partial p_1} + z\frac{\partial \psi_\theta}{\partial p_1}; (\varepsilon_{p_2p_2})^{\text{linear}} = \frac{\partial w_0}{\partial z} = 0; 2(\varepsilon_{zp_1})^{\text{linear}} \\ &= \frac{\partial w_0}{\partial p_1} + \psi_\theta; 2(\varepsilon_{p_1\theta})^{\text{linear}} \\ &= \frac{\partial v_0}{\partial p_1} + z\frac{\partial \psi_{p_1}}{\partial p_1} + \frac{1}{(R_1 - p_1\sin\alpha)}\left(\frac{\partial u_0}{\partial \theta} + z\frac{\partial \psi_\theta}{\partial \theta} + (v_0 + z\cdot\psi_{p_1})\sin\alpha\right); 2(\varepsilon_{z\theta})^{\text{linear}} \\ &= \psi_{p_1} + \frac{1}{(R_1 - p_1\sin\alpha)}\left(\frac{\partial w_0}{\partial \theta} - (v_0 + z\cdot\psi_{p_1})\cos\alpha\right); (\varepsilon_{\theta\theta})^{\text{linear}} \\ &= \frac{1}{(R_1 - p_1\sin\alpha)}\left(\frac{\partial v_0}{\partial \theta} + z\frac{\partial \psi_{p_1}}{\partial \theta} - (u_0 + z\cdot\psi_\theta)\sin\alpha + w_0\cos\alpha\right) \end{aligned} \tag{7.36}$$

$$\begin{aligned} (\varepsilon_{p_1p_1})^{\text{nonlinear}} &= \frac{1}{2}\left(\frac{\partial w_0}{\partial p_1}\right)^2; (\varepsilon_{zz})^{\text{nonlinear}} = 0; 2(\varepsilon_{zp_1})^{\text{nonlinear}} = 0; 2(\varepsilon_{p_1\theta})^{\text{nonlinear}} \\ &= \frac{1}{(R_1 - p_1\sin\alpha)}\left(\left(\frac{\partial w_0}{\partial p_1}\right)\left(\frac{\partial w_0}{\partial \theta}\right)\right); 2(\varepsilon_{z\theta})^{\text{nonlinear}} \\ &= 0; (\varepsilon_{\theta\theta})^{\text{nonlinear}} = \frac{1}{2(R_1 - p_1\sin\alpha)^2}\left(\left(\frac{\partial w_0}{\partial \theta}\right)^2 + (w_0\cos\alpha)^2\right) \end{aligned} \tag{7.37}$$

7.6.3 Nonlocal stresses of the FGM shell

According to Hooke's law $\vec{\vec{\sigma}} = C: \vec{\vec{\varepsilon}}$ (C is the material stiffness matrix), the nonlocal stresses in the spherical and conical shell structure can be obtained according to the nonlocal elasticity theory and the strain tensor introduced in the given section according to the following equations:

$$(1 - \mu\nabla^2)\sigma_{ij}^{NL} = \sigma_{ij}^{L} = C: \vec{\vec{\varepsilon}}_{ij} \quad i, j = \varphi, \theta \text{ for spherical shell and } i, j = p_1, \theta \text{ for conical shell} \tag{7.38}$$

In the modified couple stress theory, in addition to the Hooke stress, the components of the couple stress tensor $\vec{\vec{m}}$ should also be considered as $\vec{\vec{\sigma}} = C: \vec{\vec{\varepsilon}} + \vec{\vec{m}}$. We note that here, the stresses are defined locally. For the values of the components of the tensor $\vec{\vec{m}} = 2L^2G\vec{\vec{\chi}}$, the components of $\vec{\vec{\chi}}$ should be specified according to the following definition. L, G, and $\vec{\vec{\chi}}$ are the small length scale parameters, the shear modulus, and curvature tensor, respectively. $\vec{\omega}$ is the rotation vector

$$\vec{\chi} = \frac{1}{2}(\vec{\nabla}\vec{\omega} + \vec{\nabla}\,\vec{\omega}^T);\ \vec{\omega} = \frac{1}{2}\mathrm{Curl}(\vec{U}) = \frac{1}{2}\,\vec{\nabla} \times \vec{U} \tag{7.39}$$

$$\begin{aligned} &\text{Spherical shell:} \quad \vec{\omega} = \omega_\varphi \hat{e}_\varphi + \omega_\theta \hat{e}_\theta + \omega_z \hat{e}_z; \\ &\text{Conical shell:} \quad \vec{\omega} = \omega_{p_1} \hat{e}_{p_1} + \omega_\theta \hat{e}_\theta + \omega_z \hat{e}_z \end{aligned} \tag{7.40}$$

In order to avoid the readers misunderstanding the process of deriving the governing equations, only the governing equations based on the nonlocal elasticity theory will be presented in detail.

7.7 The energy method

Now, the governing equations will be obtained according to the energy method and the principle of minimum potential energy. One of the advantages of using the energy method regarding the extraction of the governing equations is that the mathematical description of the boundary conditions will also be obtained simultaneously with the extraction of the governing equations according to this approach. The total energy of the system is caused by the strain potential energy δU_ε and the energy of external forces δU_{ext}, which is introduced in the following equation

$$\delta U_{\mathrm{Total}} = \delta U_\varepsilon + \delta U_{\mathrm{ext}} = 0 \tag{7.41}$$

$$\delta U_{\mathrm{ext}} = \iint (q_z - k_w w_0 + k_p\, \nabla^2\, w_0)\delta w_0 dA \text{ For both spherical and conical shells} \tag{7.42}$$

where k_w and k_p are the Winkler and Pasternak stiffness parts of the elastic foundation. dA and dV are the area and volume differentials. The variation of strain energy and the produced energy due to the nonconservative external forces are formulated for any spherical and conical shell structure according to the following equations

$$\begin{aligned} &\text{Spherical shell: } \delta U_\varepsilon \\ &= \iiint_V \left(\sigma_{\varphi\varphi}^{NL}\delta\varepsilon_{\varphi\varphi} + \sigma_{\theta\theta}^{NL}\delta\varepsilon_{\theta\theta} + 2\sigma_{\varphi\theta}^{NL}\delta\varepsilon_{\varphi\theta} + 2\sigma_{z\varphi}^{NL}\,\delta\varepsilon_{z\varphi} + 2\sigma_{z\theta}^{NL}\delta\varepsilon_\theta\right)dV \end{aligned} \tag{7.43}$$

$$\begin{aligned} &\text{Conical shell: } \delta U_\varepsilon \\ &= \iiint_V \left(\sigma_{p_1p_1}^{NL}\,\delta\varepsilon_{p_1p_1} + \sigma_{\theta\theta}^{NL}\delta\varepsilon_{\theta\theta} + 2\sigma_{p_1\theta}^{NL}\delta\varepsilon_{p_1\theta} + 2\sigma_{zp_1}^{NL}\delta\varepsilon_{zp_1} + 2\sigma_{z\theta}^{NL}\,\delta\varepsilon_\theta\right)dV \end{aligned} \tag{7.44}$$

By integrating along the thickness z, the variation of strain potential energy for spherical and conical structures can be rewritten by considering the definition of the stress and moment resultants (N_{ij}^{NL} and M_{ij}^{NL}). Now, by collecting the same values of the displacement function variations δu_0, δv_0, δw_0, $\delta\varphi_1$, and $\delta\varphi_2$, we can obtain the governing equations of two micro- and nanoscale spherical and conical FGM spherical and conical shell structures based on the nonlocal stresses, which are introduced below.

The governing equations of micro- and nanoFGM spherical shells:

$$\delta u_0\text{: } \left(\frac{\partial N_{\varphi\varphi}^{NL}}{\partial \varphi}\right) + \cot\varphi\left(N_{\varphi\varphi}^{NL} - N_{\theta\theta}^{NL}\right) + N_{z\varphi}^{NL} + \frac{1}{\sin\varphi}\frac{\partial N_{\varphi\theta}^{NL}}{\partial \theta} = 0 \tag{7.45}$$

$$\delta v_0\text{: } \left(\frac{\partial N_{\varphi\theta}^{NL}}{\partial \varphi}\right) + 2\cot\varphi N_{\varphi\theta}^{NL} + N_{z\theta}^{NL} + \frac{1}{\sin\varphi}\frac{\partial N_{\theta\theta}^{NL}}{\partial \theta} = 0 \tag{7.46}$$

$$\begin{aligned}\delta w_0\text{: } &\frac{\partial}{\partial \varphi}\left(R\sin\varphi N_{z\varphi}^{NL}\right) - R\sin\varphi\left(N_{\varphi\varphi}^{NL} + N_{\theta\theta}^{NL}\right) + R\frac{\partial N_{z\theta}^{NL}}{\partial \theta}\\ &+ (q_z + k_w w_0 + k_p\,\nabla^2\,w_0)R^2\sin\varphi = 0\end{aligned} \tag{7.47}$$

$$\begin{aligned}\delta\psi_\theta\text{: } &R\sin\varphi M_{z\varphi}^{NL} + R^2\sin\varphi N_{z\varphi}^{NL} - \frac{\partial}{\partial\varphi}\left(R\sin\varphi M_{\varphi\varphi}^{NL}\right)\\ &+ R\sin\varphi\cot\varphi M_{\theta\theta}^{NL} - R\frac{\partial M_{\varphi\theta}^{NL}}{\partial\theta} = 0\end{aligned} \tag{7.48}$$

$$\delta\psi_\varphi\text{: } \frac{\partial M_{\varphi\theta}^{NL}}{\partial\varphi} + 2\cot\varphi M_{\varphi\theta}^{NL} + \frac{1}{\sin\varphi}\frac{\partial M_{\theta\theta}^{NL}}{\partial\theta} - \left(M_{z\theta}^{NL} + RN_{z\theta}^{NL}\right) = 0 \tag{7.49}$$

$$\begin{cases}\left(N_{\varphi\varphi}^{NL}, N_{\theta\theta}^{NL}, N_{\varphi\theta}^{NL}, N_{z\varphi}^{NL}, N_{z\theta}^{NL}\right) = \int_{-\frac{h}{2}}^{\frac{h}{2}}\left(\sigma_{\varphi\varphi}^{NL}, \sigma_{\theta\theta}^{NL}, \sigma_{\varphi\theta}^{NL}, \kappa_s\sigma_{z\varphi}^{NL}, \kappa_s\sigma_{z\theta}^{NL}\right)dz\\ \left(M_{\varphi\varphi}^{NL}, M_{\theta\theta}^{NL}, M_{\varphi\theta}^{NL}, M_{z\varphi}^{NL}, M_{z\theta}^{NL}\right) = \int_{-\frac{h}{2}}^{\frac{h}{2}}\left(\sigma_{\varphi\varphi}^{NL}, \sigma_{\theta\theta}^{NL}, \sigma_{\varphi\theta}^{NL}, \sigma_{z\varphi}^{NL}, \sigma_{z\theta}^{NL}\right)zdz\end{cases} \tag{7.50}$$

The governing equations of micro- and nanoFGM conical shells:

$$\begin{aligned}\delta u_0\text{: } &-\cos\alpha\left(\frac{\partial M_{p_1p_1}^{NL}}{\partial p_1}\right) + \sin\alpha\frac{\partial}{\partial p_1}\left(p_1 N_{p_1p_1}^{NL}\right)\\ &- R_1\left(\frac{\partial N_{p_1p_1}^{NL}}{\partial p_1}\right) - \left(\frac{\partial N_{p_1\theta}^{NL}}{\partial\theta}\right) - \sin\alpha N_{\theta\theta}^{NL} = 0\end{aligned} \tag{7.51}$$

$$\begin{aligned}\delta v_0\text{: } &-\cos\alpha\left(\frac{\partial M_{p_1\theta}^{NL}}{\partial p_1}\right) + \sin\alpha\frac{\partial}{\partial p_1}\left(p_1 N_{p_1\theta}^{NL}\right)\\ &- R_1\left(\frac{\partial N_{p_1\theta}^{NL}}{\partial p_1}\right) + \sin\alpha N_{p_1\theta}^{NL} - \cos\alpha N_{z\theta}^{NL} - \left(\frac{\partial N_{\theta\theta}^{NL}}{\partial\theta}\right) = 0\end{aligned} \tag{7.52}$$

$$\delta w_0\colon -\cos\alpha\left(\frac{\partial M_{zp_1}^{NL}}{\partial p_1}\right)+\sin\alpha\frac{\partial}{\partial p_1}\left(p_1N_{zp_1}^{NL}\right)-R_1\left(\frac{\partial N_{zp_1}^{NL}}{\partial p_1}\right)-\left(\frac{\partial N_{z\theta}^{NL}}{\partial\theta}\right)+\cos\alpha N_{\theta\theta}^{NL}$$
$$-\frac{\partial}{\partial p_1}\left((R_1-p_1\sin\alpha)N_{p_1p_1}^{NL}\left(\frac{\partial w_0}{\partial p_1}\right)\right)$$
$$-\frac{\partial}{\partial p_1}\left(M_{p_1p_1}^{NL}\cos\alpha\left(\frac{\partial w_0}{\partial p_1}\right)\right)+\frac{N_{\theta\theta}^{NL}w_0(\cos\alpha)^2}{(R_1-p_1\sin\alpha)} \tag{7.53}$$
$$-\frac{\partial}{\partial\theta}\left(\frac{N_{\theta\theta}^{NL}}{(R_1-p_1\sin\alpha)}\left(\frac{\partial w_0}{\partial\theta}\right)\right)-\left(\left(\frac{\partial w_0}{\partial\theta}\right)\left(\frac{\partial N_{p_1\theta}^{NL}}{\partial p_1}\right)+\left(\frac{\partial w_0}{\partial p_1}\right)\left(\frac{\partial N_{p_1\theta}^{NL}}{\partial\theta}\right)+2N_{p_1\theta}^{NL}\left(\frac{\partial^2 w_0}{\partial p_1\partial\theta}\right)\right)$$
$$-q_z(R_1-p_1\sin\alpha)=0$$

$$\delta\varphi\colon -\cos\alpha\left(\frac{\partial H_{p_1p_1}^{NL}}{\partial p_1}\right)+\sin\alpha\frac{\partial}{\partial p_1}\left(p_1M_{p_1p_1}^{NL}\right)-R_1\left(\frac{\partial M_{p_1p_1}^{NL}}{\partial p_1}\right)+\cos\alpha M_{zp_1}^{NL}$$
$$+(R_1-p_1\sin\alpha)N_{zp_1}^{NL}-\left(\frac{\partial M_{p_1\theta}^{NL}}{\partial\theta}\right)-\sin\alpha M_{\theta\theta}^{NL}=0 \tag{7.54}$$

$$\delta\psi\colon -\cos\alpha\left(\frac{\partial H_{p_1\theta}^{NL}}{\partial p_1}\right)+\sin\alpha\frac{\partial}{\partial p_1}\left(p_1M_{p_1\theta}^{NL}\right)-R_1\left(\frac{\partial M_{p_1\theta}^{NL}}{\partial p_1}\right)+\sin\alpha M_{p_1\theta}^{NL}$$
$$+(R_1-p_1\sin\alpha)N_{z\theta}^{NL}-\left(\frac{\partial M_{\theta\theta}^{NL}}{\partial\theta}\right)=0 \tag{7.55}$$

$$\begin{cases}\left(N_{p_1p_1}^{NL},\ N_{zp_1}^{NL},\ N_{p_1\theta}^{NL},\ N_{z\theta}^{NL},\ N_{\theta\theta}^{NL}\right)=\int_0^h\left(\sigma_{p_1p_1}^{NL},\ \kappa_s\sigma_{zp_1}^{NL},\ \sigma_{p_1\theta}^{NL},\ \kappa_s\sigma_{z\theta}^{NL},\ \sigma_{\theta\theta}^{NL}\right)dz\\ \left(M_{p_1p_1}^{NL},\ M_{zp_1}^{NL},\ M_{p_1\theta}^{NL},\ M_{z\theta}^{NL},\ M_{\theta\theta}^{NL}\right)=\int_0^h\left(\sigma_{p_1p_1}^{NL},\ \sigma_{zp_1}^{NL},\ \sigma_{p_1\theta}^{NL},\ \sigma_{z\theta}^{NL},\ \sigma_{\theta\theta}^{NL}\right)zdz\\ \left(H_{p_1p_1}^{NL},\ H_{p_1\theta}^{NL}\right)=\int_0^h\left(\sigma_{p_1p_1}^{NL},\ \sigma_{p_1\theta}^{NL}\right)z^2dz\end{cases} \tag{7.56}$$

Now, according to the definition of the nonlocal stresses based on local stresses and the nonlocal parameter $(1-\mu\nabla^2)(N,\ M)_{ij}^{NL}=(N,\ M)_{ij}^{L}$ and substituting the obtained nonlocal stresses into the governing equations, the final governing equations based on the local stress and moment resultants $(N_{ij}^L,\ M_{ij}^L)$ and nonlocal coefficient $\mu=(e_0a)^2$ are presented in the following equations

Spherical shells:

$$\delta u_0\colon \left(\frac{\partial N_{\varphi\varphi}^L}{\partial\varphi}\right)+\cot\varphi\left(N_{\varphi\varphi}^L-N_{\theta\theta}^L\right)+N_{z\varphi}^L+\frac{1}{\sin\varphi}\frac{\partial N_{\varphi\theta}^L}{\partial\theta}=0 \tag{7.57}$$

$$\delta v_0\colon \left(\frac{\partial N_{\varphi\theta}^L}{\partial\varphi}\right)+2\cot\varphi N_{\varphi\theta}^L+N_{z\theta}^L+\frac{1}{\sin\varphi}\frac{\partial N_{\theta\theta}^L}{\partial\theta}=0 \tag{7.58}$$

$$\begin{aligned}\delta w_0 &: \frac{\partial}{\partial\varphi}\left(R\sin\varphi N_{z\varphi}^{L}\right) - R\sin\varphi\left(N_{\varphi\varphi}^{L} + N_{\theta\theta}^{NL}\right) + R\frac{\partial N_{z\theta}^{L}}{\partial\theta} \\ &+ (1-\mu\nabla^2)(q_z + k_w w_0 + k_p\,\nabla^2\, w_0)R^2\sin\varphi = 0\end{aligned} \tag{7.59}$$

$$\begin{aligned}\delta\psi_\theta &: R\sin\varphi M_{z\varphi}^{L} + R^2\sin\varphi N_{z\varphi}^{L} - \frac{\partial}{\partial\varphi}\left(R\sin\varphi M_{\varphi\varphi}^{L}\right) \\ &+ R\sin\varphi\cot\varphi M_{\theta\theta}^{L} - R\frac{\partial M_{\varphi\theta}^{L}}{\partial\theta} = 0\end{aligned} \tag{7.60}$$

$$\delta\psi_\varphi: \frac{\partial M_{\varphi\theta}^{L}}{\partial\varphi} + 2\cot\varphi M_{\varphi\theta}^{L} + \frac{1}{\sin\varphi}\frac{\partial M_{\theta\theta}^{L}}{\partial\theta} - \left(M_{z\theta}^{L} + RN_{z\theta}^{L}\right) = 0 \tag{7.61}$$

Conical shells:

$$\begin{aligned}\delta u_0 &: -\cos\alpha\left(\frac{\partial M_{p_1p_1}^{L}}{\partial p_1}\right) + \sin\alpha\frac{\partial}{\partial p_1}\left(p_1 N_{p_1p_1}^{L}\right) \\ &- R_1\left(\frac{\partial N_{p_1p_1}^{L}}{\partial p_1}\right) - \left(\frac{\partial N_{p_1\theta}^{L}}{\partial\theta}\right) - \sin\alpha N_{\theta\theta}^{L} = 0\end{aligned} \tag{7.62}$$

$$\begin{aligned}\delta v_0 &: -\cos\alpha\left(\frac{\partial M_{p_1\theta}^{L}}{\partial p_1}\right) + \sin\alpha\frac{\partial}{\partial p_1}\left(p_1 N_{p_1\theta}^{L}\right) \\ &- R_1\left(\frac{\partial N_{p_1\theta}^{L}}{\partial p_1}\right) + \sin\alpha N_{p_1\theta}^{L} - \cos\alpha N_{z\theta}^{L} - \left(\frac{\partial N_{\theta\theta}^{L}}{\partial\theta}\right) = 0\end{aligned} \tag{7.63}$$

$$\begin{aligned}&\delta w_0: -\cos\alpha\left(\frac{\partial M_{zp_1}^{L}}{\partial p_1}\right) + \sin\alpha\frac{\partial}{\partial p_1}\left(p_1 N_{zp_1}^{L}\right) - R_1\left(\frac{\partial N_{zp_1}^{L}}{\partial p_1}\right) - \left(\frac{\partial N_{z\theta}^{L}}{\partial\theta}\right) + \cos\alpha N_{\theta\theta}^{L} \\ &- (1-\mu\nabla^2)\left(\frac{\partial}{\partial p_1}\left((R_1 - p_1\sin\alpha)N_{p_1p_1}^{NL}\left(\frac{\partial w_0}{\partial p_1}\right)\right) - \frac{\partial}{\partial p_1}\left(M_{p_1p_1}^{NL}\cos\alpha\left(\frac{\partial w_0}{\partial p_1}\right)\right)\right. \\ &+ \frac{N_{\theta\theta}^{NL}w_0(\cos\alpha)^2}{(R_1 - p_1\sin\alpha)} - \frac{\partial}{\partial\theta}\left(\frac{N_{\theta\theta}^{NL}}{(R_1 - p_1\sin\alpha)}\left(\frac{\partial w_0}{\partial\theta}\right)\right) - \left(\left(\frac{\partial w_0}{\partial\theta}\right)\left(\frac{\partial N_{p_1\theta}^{NL}}{\partial p_1}\right) + \left(\frac{\partial w_0}{\partial p_1}\right)\left(\frac{\partial N_{p_1\theta}^{NL}}{\partial\theta}\right) + 2N_{p_1\theta}^{NL}\left(\frac{\partial^2 w_0}{\partial p_1\partial\theta}\right)\right) \\ &\left. - q_z(R_1 - p_1\sin\alpha)\right) = 0\end{aligned} \tag{7.64}$$

$$\begin{aligned}\delta\varphi &: -\cos\alpha\left(\frac{\partial H_{p_1p_1}^{L}}{\partial p_1}\right) + \sin\alpha\frac{\partial}{\partial p_1}\left(p_1 M_{p_1p_1}^{L}\right) - R_1\left(\frac{\partial M_{p_1p_1}^{L}}{\partial p_1}\right) + \cos\alpha M_{zp_1}^{L} \\ &+ (R_1 - p_1\sin\alpha)N_{zp_1}^{L} - \left(\frac{\partial M_{p_1\theta}^{L}}{\partial\theta}\right) - \sin\alpha M_{\theta\theta}^{L} = 0\end{aligned} \tag{7.65}$$

$$
\begin{aligned}
\delta\psi\colon &-\cos\alpha\left(\frac{\partial H^L_{p_1\theta}}{\partial p_1}\right)+\sin\alpha\frac{\partial}{\partial p_1}\left(p_1 M^L_{p_1\theta}\right)-R_1\left(\frac{\partial M^L_{p_1\theta}}{\partial p_1}\right)+\sin\alpha M^L_{p_1\theta}\\
&+(R_1-p_1\sin\alpha)N^L_{z\theta}-\left(\frac{\partial M^L_{\theta\theta}}{\partial\theta}\right)=0
\end{aligned}
\tag{7.66}
$$

7.8 The boundary conditions and solution method

The mathematical formulation of the boundary conditions for each shell can be obtained from the first part of the integration by parts when deriving the governing equations using the energy method. Also, the solution method used in this chapter for solving the partial differential equations, is the SAPM method, and the details of this method can be found in previous studies [43–45]. There are three sets of boundary conditions in this chapter: clamped (*C*), supported (*S*), and free edges (*F*). Their equations are expressed below. The conical shell:

$$
\begin{cases}
C\colon (u_0=v_0=w_0=\psi_{p_1}=\psi_\theta=0; p_1=0,\ L \text{ and } \theta=0,\ \tau)\\
S\colon (u_0=v_0=w_0=0; p_1=0,\ L \text{ and } \theta=0,\ \tau),\ (M_{p_1p_1}=\psi_\theta=0; p_1=0,\ L),\\
\qquad (M_{\theta\theta}=\psi_{p_1}=0; \theta=0,\ \tau)\\
F\colon (N_{p_1p_1}=N_{p_1\theta}=M_{p_1\theta}=0; p_1=0,\ L \text{ and } \theta=0,\ \tau),\ (N_{p_1p_2}=M_{p_1p_1}=0; p_1=0,\ L)\\
\qquad (N_{p_2\theta}=M_{\theta\theta}=0; \theta=0,\ \tau)
\end{cases}
\tag{7.67}
$$

The spherical shell:

$$
\begin{cases}
C\colon (u_0=v_0=w_0=\psi_\varphi=\psi_\theta=0; \varphi=\varphi_1,\ \varphi_2 \text{ and } \theta=0,\ \tau)\\
S\colon (u_0=v_0=w_0=0; \varphi=\varphi_1,\ \varphi_2 \text{ and } \theta=0,\ \tau),\ (M_{\varphi\varphi}=\psi_\theta=0; \varphi=\varphi_1,\ \varphi_2),\\
\qquad (M_{\theta\theta}=\psi_\varphi=0; \theta=0,\ \tau)\\
F\colon (N_{p_1p_1}=N_{p_1\theta}=M_{p_1\theta}=0; p_1=0,\ L \text{ and } \theta=0,\ \tau),\ (N_{z\varphi}=M_{\varphi\varphi}=0; p_1=0,\ L)\\
\qquad (N_{z\theta}=M_{\theta\theta}=0; \theta=0,\ \tau)
\end{cases}
\tag{7.68}
$$

7.9 Numerical results

7.9.1 Validation

To compare the obtained results by the presented analysis and the ABAQUS finite element software and to ensure the accuracy of the results, two figures 7.3 and 7.4 of the deflection changes for two spherical and conical shell structures are presented. The material and geometric specifications of the two shell structures are given below. It is impossible to analyze micro- and nanoscale structures in finite element software such as ABAQUS. Therefore, macroscale shells have been considered to evaluate the results

$$
\begin{aligned}
&\text{Spherical: } E_c=185\text{ GPa};\ E_m=190\text{ GPa};\ \nu_c=\nu_m=0.29;\ g=0;\ h=5\text{ mm};\\
&\varphi_1=20^\circ;\ \varphi_2=90^\circ;\ \theta_1=0;\ \varphi_2=180^\circ;\ q_z=10\text{ MPa}
\end{aligned}
\tag{7.69}
$$

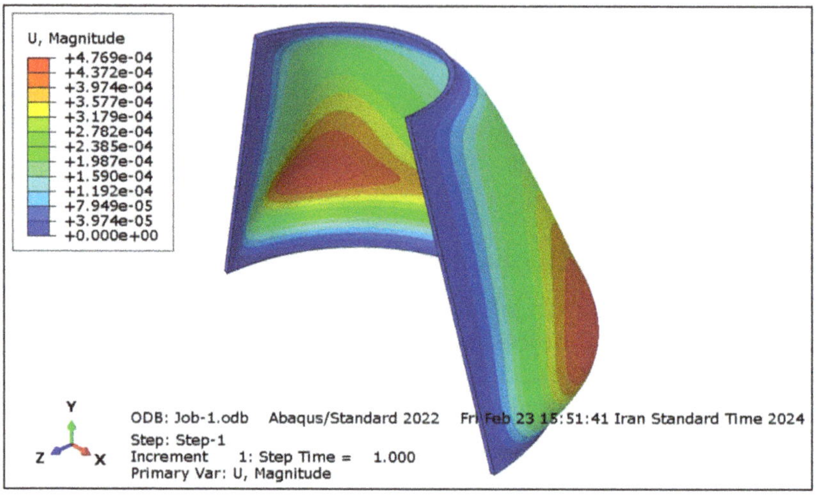

(a)

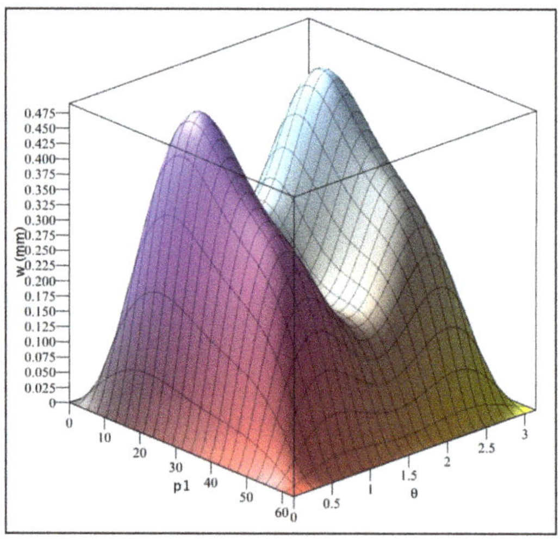

(b)

Figure 7.3. Comparison between the results of (a) present analysis and (b) ABAQUS for conical shell structure.

$$\begin{aligned}&\text{Conical:}\quad E_c = 190\ \text{GPa};\ E_m = 190\ \text{GPa};\ \nu_c = \nu_m = 0.29;\ g = 0;\ h = 5\ \text{mm};\\ &R_1 = 0.2\ \text{m};\ R_2 = 0.1\ \text{m};\ H = 0.3\ \text{m};\ \theta_1 = 0;\ \theta_2 = 180^{\circ};\ q_z = 10\ \text{MPa}\end{aligned} \tag{7.70}$$

As can be seen, the results from the presented static equilibrium equations show a very good agreement with the results of the ABAQUS software. Therefore, the simulation performed in this chapter can be used with complete confidence regarding the static analysis of spherical and conical shell structures.

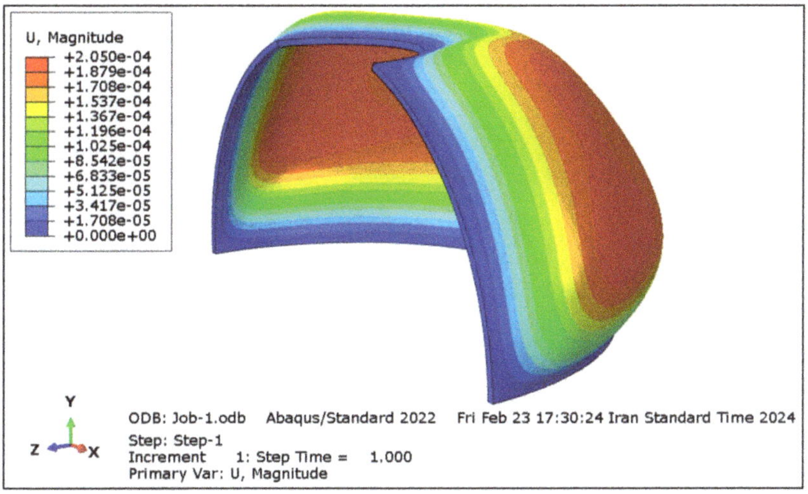

(a)

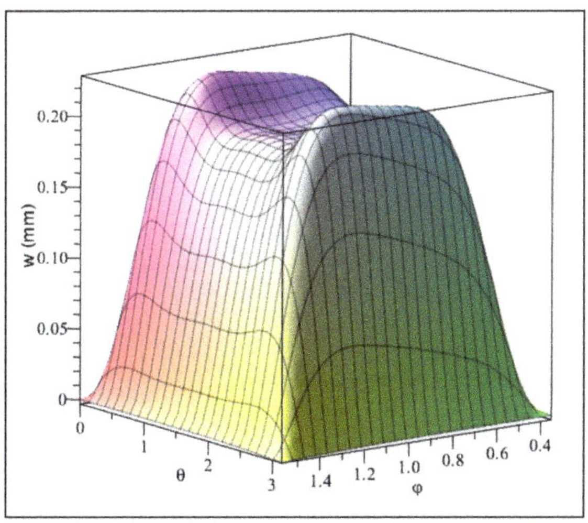

(b)

Figure 7.4. Comparison between the results of (a) present analysis and (b) ABAQUS for spherical shell structure.

7.9.2 Nonlocal and FGM numerical results

One of the most critical parameters determining the functional characteristics of FGM materials is the intensity of changes in material properties from metal to ceramic g. Figure 7.5 is drawn to investigate the effect of the g parameter value on the deflection of a conical and spherical shell structure. As can be seen at the beginning, by choosing $g = 0$, all the properties of the material through the thickness are pure metal, and by increasing the value of g, the properties of

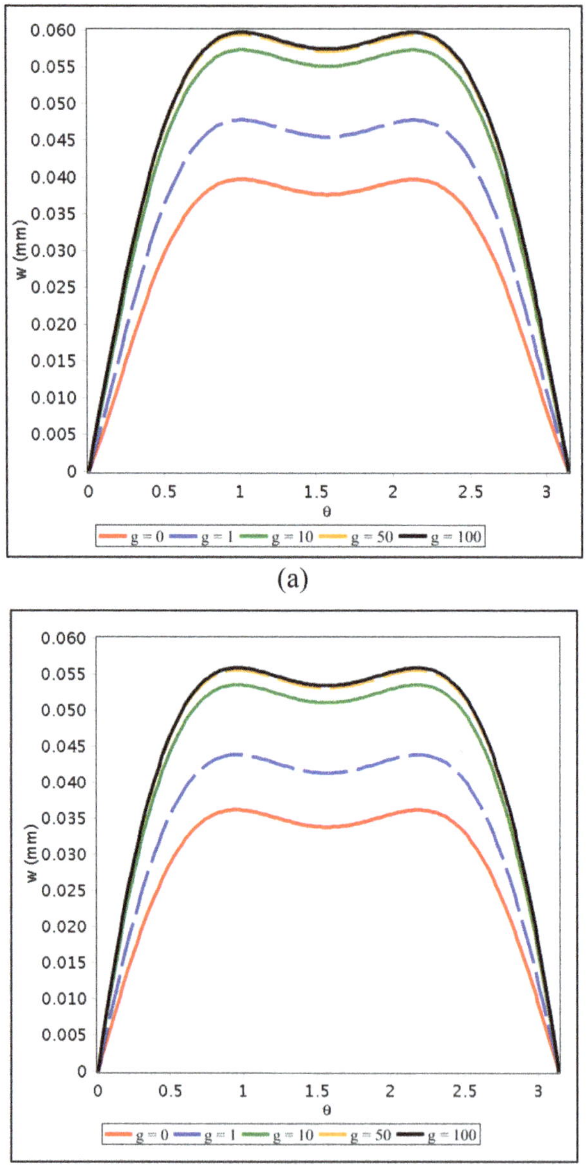

Figure 7.5. Deflection changes versus g parameter of FGM material for (a) conical and (b) spherical shell structure.

the metal decrease with a sharp slope, and the ceramic properties are increased at the beginning. However, the increase of g will have a minimal effect on the variation rate in the deformation. Figure 7.5 provides appropriate information regarding the amount of metal or ceramic property of FGM material according to the changes of parameter g. This issue will lead to the optimal selection of a structure when the desired design goal is, for example, that a structure's surface tends to have more ceramic properties like the surfaces where we should have proper insulation against heat and metal properties on the other.

As mentioned earlier in the geometry description, one of the objectives of choosing the analysis of the conical shell structure is to obtain a simulation for the cylindrical shell structure according to the obtained equations for the conical shape. According to figure 7.1, if the value of angle α is chosen equal to zero, the cone will turn into a cylinder, and there is no need to simulate a cylindrical shell and rewrite the equations separately. Therefore, this approach will also present the static results for the cylindrical FGM shell.

Figure 7.6 investigates the influence of the nonlocal parameter (e_0a) and the power-law index of the FGM material (g), for three conical, spherical, and cylindrical shells. Each analyzed shell structure's material and geometric properties (in figure 7.6) are presented below

$$\begin{aligned}&\text{Conical: } E_c = 185\text{ GPa}; E_m = 70\text{ GPa}; \nu_c = \nu_m = 0.3;\ h = 0.7\text{ nm};\\ &R_1 = 10\text{ nm}; R_2 = 5\text{ nm}; H = 20\text{ nm}; \theta_1 = 0; \theta_2 = 180^\circ; q_z = 0.1\text{ GPa}\end{aligned} \tag{7.71}$$

$$\begin{aligned}&\text{Spherical: } E_c = 185\text{ GPa}; E_m = 70\text{ GPa}; \nu_c = \nu_m = 0.3; h = 0.7\text{ nm};\\ &R = 12\text{ nm}; \varphi_1 = 20^\circ; \varphi_2 = 90^\circ; \theta_1 = 0; \varphi_2 = 180^\circ; q_z = 0.1\text{ GPa}\end{aligned} \tag{7.72}$$

$$\begin{aligned}&\text{Cylindrical: } E_c = 185\text{ GPa}; E_m = 70\text{ GPa}; \nu_c = \nu_m = 0.3; h = 0.7\text{ nm};\\ &R = 5\text{ nm}; H = 20\text{ nm}; \theta_1 = 0;\ \theta_2 = 180^\circ;\ q_z = 0.1\text{ GPa}\end{aligned} \tag{7.73}$$

Each figure is drawn for two values of g and e_0a. It can be seen that with the increase of the nonlocal parameter, the resulting static deformations decrease in all three structures. Also, for example, in a conical shell, if the nonlocal parameter $e_0a = 2$ nm and $g = 1$, the deformations in the longitudinal direction will decrease steeply at first, and then after a maximum, it can be observed that if $e_0a = 4$ nm and the parameter g does not change, the maximum value of the deflection will tend towards the center of the cone length. Therefore, in the nonlocal analysis, when the small scale is considered, the behavior of the FGM shell structure is different from the classical analysis, and the results of the classical analysis will not be acceptable. Another result from figure 7.6 is that for different values of e_0a and g equal deformations can be obtained for the FGM shell structure under the same boundary conditions and uniform loading. For example, in two conical and cylindrical structures, similar deformations can be seen for the values of $g = 1$; $e_0a = 2$ nm and $g = 5$; $e_0a = 4$ nm. The explanation mentioned above is also true for the spherical shell structure. Therefore, proper control of the effective parameters g and e_0a can obtain similar deformation values. Figure 7.6 provides excellent information about the effect of two important parameters related to the small-scale e_0a and the power-law index of the FGM material.

One of the issues discussed in this chapter is the structural defect of porosity. Porosity will reduce the structure's resistance, which can be considered a factor in

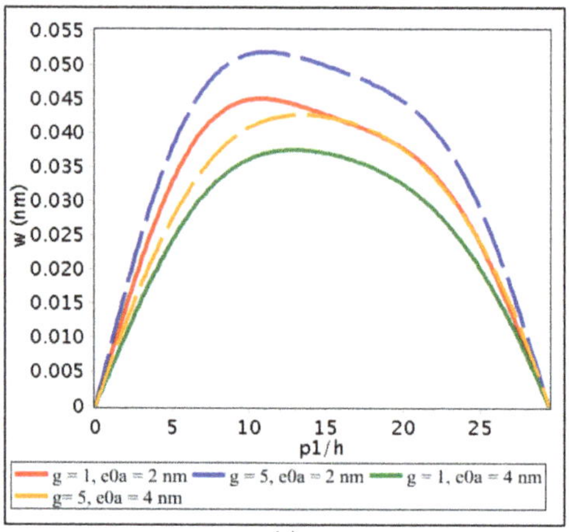

(a)

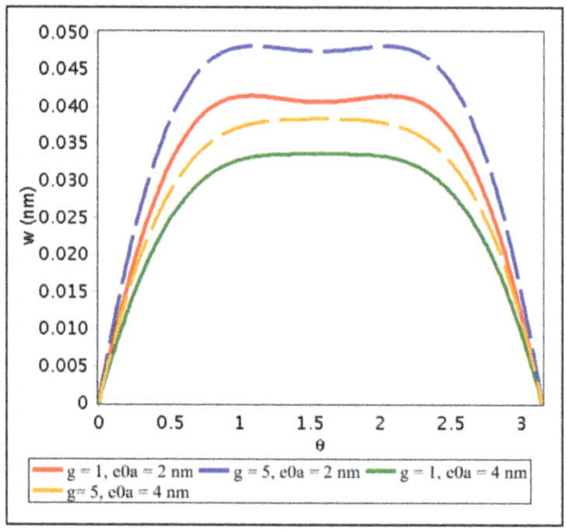

(b)

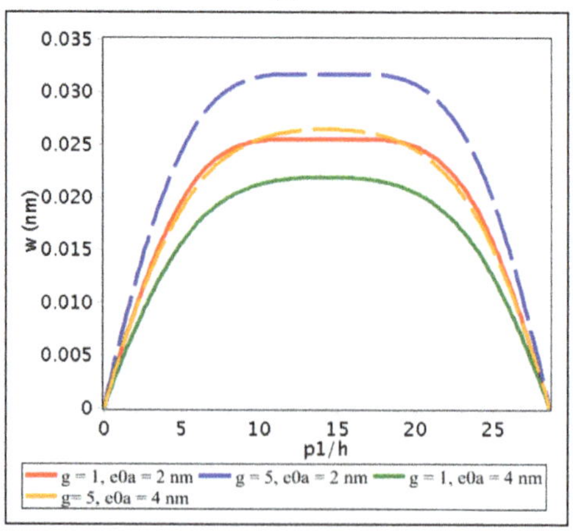

(c)

Figure 7.6. Deformation changes for different amounts of the nonlocal parameter e_0a and the power-law index g for (a) conical, (b) spherical, and (c) cylindrical FGM shell structure.

reducing Young's elasticity modulus in two ways (even and uneven) with the following equations:

$$\text{Even: } E(z) = (E_c - E_m)\left(\frac{z}{h} + \frac{1}{2}\right)^g + E_m - \frac{\lambda}{2}(E_c + E_m)\left(-\frac{h}{2} \leqslant z \leqslant \frac{h}{2}\right) \tag{7.74}$$

$$\begin{aligned}\text{Uneven: } E(z) = {} & (E_c - E_m)\left(\frac{z}{h} + \frac{1}{2}\right)^g + E_m \\ & - \frac{\lambda}{2}(E_c + E_m)\left(1 - \frac{2\,|z|}{h}\right)\left(-\frac{h}{2} \leqslant z \leqslant \frac{h}{2}\right)\end{aligned} \tag{7.75}$$

In the above equations, λ is a dimensionless number between zero and one, which determines the porosity value from the lowest value of zero to the highest value, $\lambda = 1$. Figure 7.7 shows the effect of the λ coefficient for two structural defect modes, even and uneven (the same properties in figure 7.6). In the case of uneven structural defect distribution, with the increase of the λ factor, the structural deformation changes will be minor compared to the even case. Figure 7.8 is drawn with the same geometric and material characteristics as figure 7.7, with the difference that the parameter g of the FGM material has changed from the value of $g = 1$ to $g = 100$. It can be seen that with the increase of parameter g, the material's properties have become metal. In this case, the deflection changes in the length of the conical shell will be more uniform, and with the increase of the λ parameter, we will see a greater distance in the deformation of the structure. Figure 7.9 is also drawn with the same conditions as figure 7.7; however, the nonlocal coefficient is increased to $e_0a = 4$ nm. According to figure 7.9, the noteworthy point is that the deformations become more uniform with the increase of the porosity coefficient. Figures 7.7–7.9 provide appropriate information regarding the effects of the nonlocal coefficient of small-scale analysis, the power-law parameter of the FGM material, and the porosity value on the deformation results of the FGM conical shell nanostructure.

7.10 Conclusion and remarks

This chapter has investigated the static bending analysis of micro- and nanoFGM shells. In order to simulate the characteristics of the small-scale structure, the modified couple stress theory was briefly introduced, and Eringen's nonlocal elasticity theory was used to obtain the static governing equations of the conical and spherical FGM shells. The governing equations and the mathematical description of the boundary conditions were obtained using the energy method. According to the geometric shape of the conical shell, the results related to cylindrical shells can be easily obtained. The obtained partial differential equations were solved using the SAPM solution method. Some of the achievements of the current chapter can be categorized and introduced as follows:

- The nonlinear bending results for conical and spherical shells obtained by the governing equations in this chapter agree with the results obtained from the ABAQUS finite element software.

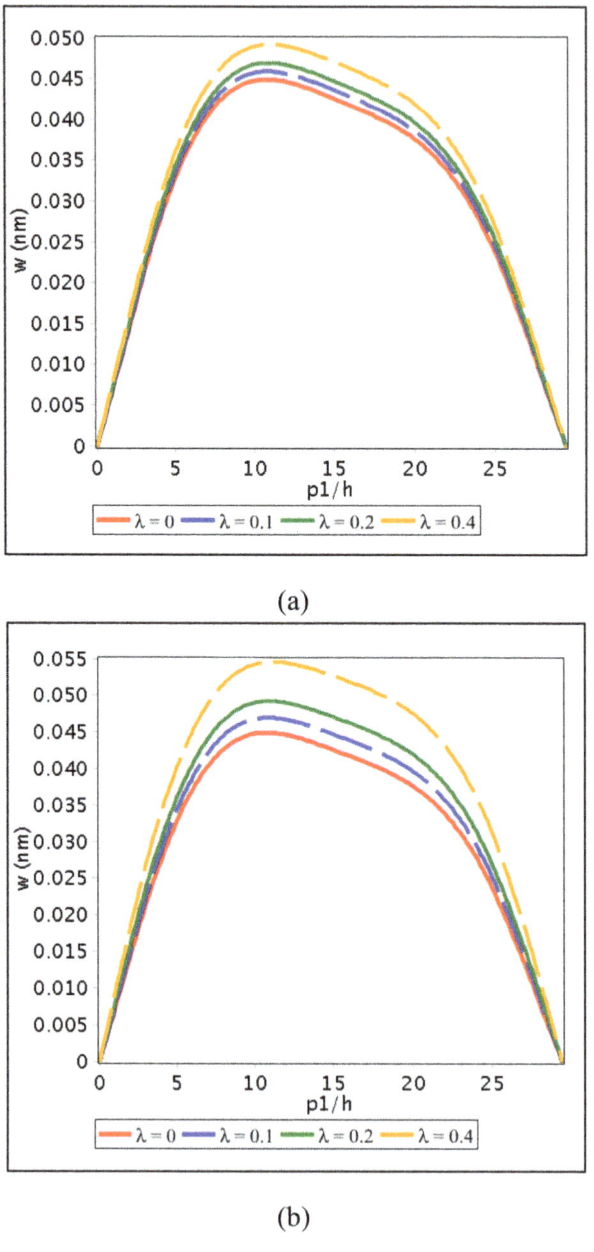

Figure 7.7. Deformation variations of a conical FGM nanoshell ($g = 1$ and $e_0a = 2$ nm) for (a) uneven and (b) even porosities.

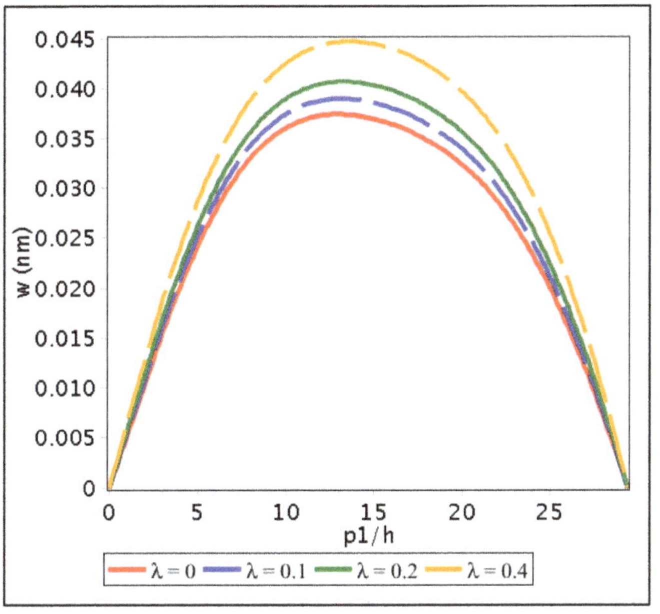

Figure 7.8. Deflection changes with similar properties in figure 7.7(b) and $g = 100$ and $e_0a = 2$ nm.

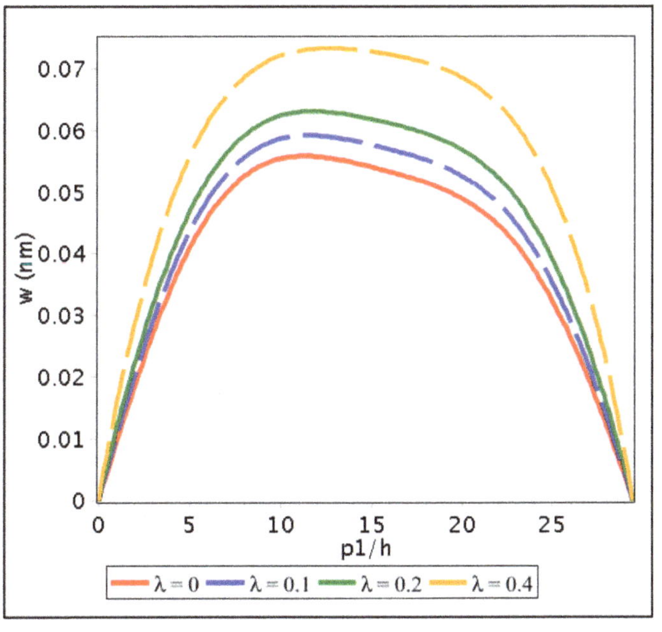

Figure 7.9. Deflection changes with similar properties in figure 7.7(b) and $g = 1$ and $e_0a = 4$ nm.

- According to the equations introduced in this chapter, different types of boundary conditions, such as clamped, simply supported, and free edge, can be considered.
- The increase in the value of FGM's power-law index (g) at the beginning of the changes causes the FGM's properties to change from metal to ceramic with a steeper slope. As the g parameter continues to increase, the change process will decrease.
- Increasing the nonlocal coefficient in the FGM nanoshell analysis will reduce the structure's deformation.
- Increasing the size-dependent parameter in the cone-shaped shell structure will reduce the extreme deflection changes in the cone's longitudinal direction.
- The effect of the nonlocal coefficient on the deformation results of the FGM shell structure is greater than the power-law coefficient of the FGM.
- The even type of porosity will have a more significant effect on reducing the strength of the FGM nanoshell structure. Also, the value of the nonlocal parameter in the small-scale analysis and the characteristic parameter g in the analyzed FGM nanoshell material will intensify the effect of porosity.

References

[1] Weikai L and Baohong H 2018 Research and application of functionally gradient materials *IOP Conf. Ser.: Mater. Sci. Eng.* **394** 022065

[2] Benferhat R, Daouadji T H and Mansour M S 2017 A review on functionally gradient materials (FGMs) and their applications *IOP Conf. Ser.: Mater. Sci. Eng.* **229** 012021

[3] Mahamood R M and Akinlabi E T 2020 Types of Functionally Graded Materials and Their Areas of Application *Mining, Metallurgy and Materials Engineering Electronic* (Berlin: Springer) https://doi.org/10.1007/978-3-319-53756-6_2

[4] El-Galy I M, Saleh B I and Ahmed M H 2019 Functionally graded materials classifications and development trends from industrial point of view *SN Appl. Sci.* **1** 1378

[5] Zhang N, Khan T, Guo H, Shi S, Zhong W and Zhang. W 2019 Functionally graded materials: an overview of stability, buckling, and free vibration analysis *Adv. Mater. Sci. Eng.* **2019** 1354150

[6] Weiss L, Nessler Y, Novelli M, Laheurte P and Grosdidier T 2019 On the use of functionally graded materials to differentiate the effects of surface severe plastic deformation, roughness and chemical composition on cell proliferation *Metals* **9** 1344

[7] Baghlani A, Khayat M and Dehghan S M 2020 Free vibration analysis of FGM cylindrical shells surrounded by Pasternak elastic foundation in thermal environment considering fluid-structure interaction *Appl. Math. Modelling* **78** 550–75

[8] Kim Y W 2015 Free vibration analysis of FGM cylindrical shell partially resting on Pasternak elastic foundation with an oblique edge *Composites* B **70** 263–76

[9] Bagheri H, Kiani Y, Bagheri N and Eslami M R 2020 Free vibration of joined cylindrical–hemispherical FGM shells *Arch. Appl. Mech.* **90** 2185–99

[10] Karroubi R and Irani-Rahaghi M 2019 Rotating sandwich cylindrical shells with an FGM core and two FGPM layers: free vibration analysis *Compos. B Eng.* **40** 563–78

[11] Naeem M N, Khan A G, Arshad S H, Shah A G and Gamkhar M 2014 Vibration of three-layered FGM cylindrical shells with middle layer of isotropic material for various boundary conditions *World J. Mech.* **4** 315–31

[12] Shahmohammadi M A, Azhari M and Saadatpour M M 2020 Free vibration analysis of sandwich FGM shells using isogeometric B-spline finite strip method *Steel Compos. Struct.* **34** 361–76
[13] Fakhari Golpayegani I, Mohammadi Arani E and Foroughifar A A 2019 Finite element vibration analysis of variable thickness thin cylindrical FGM shells under various boundary conditions *Mater. Perform. Charact.* **8** 491–502
[14] Shahmohammadi M A, Azhari M, Saadatpour M M and Sarrami-Foroushani. S 2020 Stability of laminated composite and sandwich FGM shells using a novel isogeometric finite strip method *Eng. Comput.* **37** 1369–95
[15] Amieur B, Djermane M and Hammadi F 2017 Nonlinear analysis of degenerated FGM shells *Int. J. Appl. Eng. Res.* **12** 11511–22
[16] Esmaeili H R, Arvin H and Kiani Y 2019 Axisymmetric nonlinear rapid heating of FGM cylindrical shells *J. Therm. Stresses* **42** 490–505
[17] Hajlaoui A, Chebbi E, Wali M and Dammak F 2020 Geometrically nonlinear analysis of FGM shells using solidshell element with parabolic shear strain distribution *Int. J. Mech. Mater. Des.* **16** 351–66
[18] Shen H S 2007 Thermal postbuckling of shear deformable FGM cylindrical shells with temperature-dependent properties *Mech. Adv. Mater. Struct.* **14** 439–52
[19] Sofiyev A H 2011 On the vibration and stability of clamped FGM conical shells under external loads *J. Compos. Mater.* **45** 771
[20] Dastjerdi S and Akgöz B 2018 New static and dynamic analyses of macro and nano FGM plates using exact three-dimensional elasticity in thermal environment *Compos. Struct.* **192** 626–41
[21] Dastjerdi S, Akgöz B and Civalek. Ö 2020 On the effect of viscoelasticity on behavior of gyroscopes *Int. J. Eng. Sci.* **149** 103236
[22] Dastjerdi S, Malikan M, Eremeyev V A, Akgöz B and Civalek Ö 2021 On the generalized model of shell structures with functional cross-sections *Compos. Struct.* **272** 114192
[23] Dastjerdi S, Malikan M, Dimitri R and Tornabene F 2021 Nonlocal elasticity analysis of moderately thick porous functionally graded plates in a hygro thermal environment *Compos. Struct.* **255** 112925
[24] Ghareghani S, Loghman A and Mohammadimehr M 2021 Analysis of FGM micro cylindrical shell with variable thickness using Cooper Naghdi model: bending and buckling solutions *Mech. Res. Commun.* **115** 103739
[25] Sladek V, Sladek J, Repka M and Sator L 2020 FGM micro/nano-plates within modified couple stress elasticity *Compos. Struct.* **245** 112294
[26] Khorshidi S, Chakouvari S, Askari H and Cveticanin L 2022 Free vibrations of flexoelectric FGM conical nanoshells with piezoelectric layers: modeling and analysis *Energies* **15** 2973
[27] Sahmani S and Fattahi A M 2018 Small scale effects on buckling and postbuckling behaviors of axially loaded FGM nanoshells based on nonlocal strain gradient elasticity theory *Appl. Math. Mech.* **39** 561–80
[28] Yang X, Sahmani S and Safaei B 2021 Postbuckling analysis of hydrostatic pressurized FGM microsized shells including strain gradient and stress-driven nonlocal effects *Eng. Comput.* **37** 1549–64
[29] Shojaeefard M H, Saeidi Googarchin H, Mahinzare M and Adibi M 2018 Vibration and buckling analysis of a rotary functionally graded piezomagnetic nanoshell embedded in viscoelastic media *J. Intell. Mater. Syst. Struct.* **29** 2344–61

[30] Ma X, Sahmani S and Safaei B 2022 Quasi-3D large deflection nonlinear analysis of isogeometric FGM microplates with variable thickness via nonlocal stress–strain gradient elasticity *Eng. Comput.* **38** 3691–704

[31] Sahmani S and Mohammadi Aghdam M 2017 Imperfection sensitivity of the size dependent postbuckling response of pressurized FGM nanoshells in thermal environments *Arch. Civ. Mech. Eng.* **17** 623–38

[32] Shahzad M A, Safaei B, Sahmani S, Basingab M S and Hameed A Z 2023 Nonlinear three-dimensional stability characteristics of geometrically imperfect nanoshells under axial compression and surface residual stress *Nanotechnol. Rev.* **12** 20220551

[33] Timesli A 2022 Analytical modeling of buckling behavior of porous FGM cylindrical shell embedded within an elastic foundation *Gazi Univ. J. Sci.* **35** 148–65

[34] Sheykhi A, Hosseini-Hashemi S, Maghsoudpour A and Etemadi Haghighi S 2022 Free nonlinear vibration analysis of nano-truncated conical shells based on modified strain gradient theory *Proc. Inst. Mech. Eng., Part* L **236** 110–46

[35] Hadi A, Zamani Nejad M, Rastgoo A and Hosseini M 2018 Buckling analysis of FGM Euler–Bernoulli nano-beams with 3D-varying properties based on consistent couple-stress theory *Steel Compos. Struct.* **26** 663–72

[36] Jung W Y and Han S C 2013 Analysis of sigmoid functionally graded material (S-FGM) nanoscale plates using the nonlocal elasticity theory *Math. Probl. Eng.* **2013** 476131

[37] Ghandourah E E, Amine Daikh A, Khatir S, Alhawsawi A M, Banoqitah E M and Eltaher. M A 2023 A dynamic analysis of porous coated functionally graded nanoshells rested on viscoelastic medium *Mathematics* **11** 2407

[38] Shariati M, Shishehsaz M and Mosalmani R 2023 Stress-driven approach to vibrational analysis of FGM annular nano-plate based on first-order shear deformation plate theory *J. Appl. Comput. Mech.* **9** 637–55

[39] Mongea J C and Mantari J L 2020 3D elasticity numerical solution for the static behavior of FGM shells *Eng. Struct.* **208** 110159

[40] Lori Dehsaraji M, Arefi M and Loghman A 2021 Size dependent free vibration analysis of functionally graded piezoelectric micro/nano shell based on modified couple stress theory with considering thickness stretching effect *Def. Technol.* **17** 119–34

[41] Hosseini-Hashemi S, Sharifpour F and Ilkhani M R 2016 On the free vibrations of size dependent closed micro/nano spherical shell based on the modified couple stress theory *Int. J. Mech. Sci.* 115–6 501–15

[42] Gholami R, Darvizeh A, Ansari R and Sadeghi F 2016 Vibration and buckling of first-order shear deformable circular cylindrical micro-/nanoshells based on Mindlin's strain gradient elasticity theory *Eur. J. Mech. A. Solids* **58** 76–88

[43] Dastjerdi S, Alibakhshi A, Akgöz B and Civalek Ö 2022 A novel nonlinear elasticity approach for analysis of nonlinear and hyperelastic structures *Eng. Anal. Boundary Elem.* **143** 219–36

[44] Dastjerdi S, Alibakhshi A, Akgöz B and Civalek Ö 2023 On a comprehensive analysis for mechanical problems of spherical structures *Int. J. Eng. Sci.* **183** 103796

[45] Dastjerdi S and Jabbarzadeh M 2017 Non-linear bending analysis of multi-layer orthotropic annular/circular graphene sheets embedded in elastic matrix in thermal environment based on non-local elasticity theory *Appl. Math. Modelling* **41** 83–101

IOP Publishing

Advances in Modeling and Analysis of Functionally Graded Micro- and Nanostructures

Subrat Kumar Jena, S Pradyumna and S Chakraverty

Chapter 8

Applications of functionally graded nano- and microstructures in MEMS and NEMS

Vikash Kumar, Naveen Kumar Akkasali, Erukala Kalyan Kumar and Subrata Kumar Panda

Functionally graded nano- and microstructures (FGNMS) have emerged as vital components in microelectromechanical systems (MEMS) and nanoelectromechanical systems (NEMS), providing unparalleled adaptability and functionality. This chapter outlines the numerous applications and benefits of FGNMS in MEMS and NEMS devices. FGNMS offers customizable mechanical, thermal, electrical, and optical functions by combining components with different compositions and properties at the nano- and microscale, improving device performance and eliminating existing constraints. This chapter addresses major applications for FGNMS, including sensing, actuation, energy harvesting, and biomedical devices, where they outperform in terms of sensitiveness, performance, and miniaturization. Furthermore, it emphasizes the fabrication procedures and design factors that are critical to the effective application of FGNMS in MEMS and NEMS devices. A comprehensive review highlights the transformational potential, capabilities, and applications of FGNMS in MEMS and NEMS structures.

8.1 Introduction

Microelectromechanical systems (MEMS) and nanoelectromechanical systems (NEMS) are cutting-edge technologies that have transformed several industries such as electronics, healthcare, and aerospace. The incorporation of functionally graded (FG) nano and microstructures has increased the capabilities and performance of these systems. This chapter addresses the uses of FG nano and microstructures in MEMS and NEMS, emphasizing their importance in stretching technical boundaries.

doi:10.1088/978-0-7503-6024-1ch8

The mechanical/electrostatic responses of FG MEMS/NEMS electrode/ microbeam structure has been analyzed analytically under the Winkler–Pasternak foundation using the galerkin or shooting method utilizing Euler–Bernoulli theory. The study shows that the grading of the functionally graded material (FGM) produces variation in the pull-in voltage [1, 2]. The pull-in behavior of the cantilever beam has been reported of micro- and nanoFGM beams with small-scale electro- static force effect. The study concludes that a micro/nanobeam model based on the FGM structure performs stiffer and has higher pull-in voltages [3]. The dynamic instability of an electrically actuated FG nanobeam has been testified using the Casimir attraction model. The effects of voltage and graded power in FGM on the dynamic pull-in voltage and pull-in duration of vibrating nanoactuators are addressed in Sedighi *et al* [4]. FG carbon nanotubes' dynamic pull-in instability-reinforced nanoactuators with damping behavior has been analyzed utilizing van der Waals interaction and Casimir force. It is noticed that the dynamic pull-in voltage of the system with damping behavior is bigger than that without a damping system since the dissipative impact of damping demands additional power to be pumped into the system [5]. The stability responses of electrically actuated/piezoelectric FG nanoshells are analyzed using shear deformable shell theory based on Hamilton's principle and the differential quadrature method. The outcomes show that increased electric voltage and temperature lead to lower frequency values for circular FG nanoplates [6–8].

Previous investigations have identified the application of FGNMS in MEMS and NEMS. Some research suggests that improving grading patterns and material in FGNMS leads to improved system performance and reliability. This chapter highlights several applications and benefits related to FGNMS in MEMS and NEMS devices. FGNMS provides customizable mechanical, thermal, electrical, and optical functionalities by integrating components with varying compositions and qualities at the nano- and microscale, enhancing device performance while reducing existing stipulations. Additionally, it focuses on the fabrication techniques and design elements crucial to the successful implementation of FGNMS in MEMS and NEMS systems.

8.2 Fundamentals of functionally graded nano- and microstructures

8.2.1 Definition and characteristics

The latest developments in manufacturing techniques, such as powder metallurgy, have made it possible to modify a structure's mechanical properties on a micro/ nanoscale [9]. This is achieved by creating structures from a layered blend of two or more materials, known as FG structures [10]. FG structures have spatially varying mechanical and electrical properties, making them helpful in improving the performance of many microelectromechanical and nanoelectromechanical systems.

Characteristics

- The FG (micro- and nanoscale) materials exhibit alterations in material properties, typically in at least one direction [10].

- FGM replaces sharp transitions in material behavior with smooth transitions and continuous variation of material properties such as Young's modulus, Poisson's ratio, density, shear modulus, and thermal expansion in the desired direction [11].
- The FG material property profiles are adjustable to achieve the desirable mechanical characteristics of micro- and nanoscale structures [10].
- The gradual change in micro- and nanoscale FG material volume fraction and the nonidentical structure in the preferred direction result in continuously graded properties, such as thermal conductivity, corrosion resistance, specific heat, hardness, and stiffness ratio [11].

8.2.2 Fabrication techniques

The production of FG material at micro- and nanoscale encompasses a variety of methods, including centrifugal casting, powder metallurgy, plasma spraying, chemical and physical vapor deposition (CVD/PVD), lamination and infiltration methods, and solid freeform fabrication (SFF) or additive manufacturing (AM). The detailed steps are enumerated in the subsequent sections.

8.2.2.1 Centrifugal casting method

Centrifugal casting is used to fabricate FGNMS at the micro- and nanoscale, as shown in figure 8.1. The method uses the radial stresses created by centrifugal action to separate the unique second phase from the materials [12]. When centrifugal force is applied to a slurry containing particles, it forms two different regions: one augmented in particles and one depleted in particles. The level of particle separation and the specific positions of the augmented and depleted particle zones within the casting depend mainly on factors like melt temperature, metal viscosity, cooling rate, densities of particles and liquid, particle size, and the level of centrifugal acceleration [13]. The behavior of particles varies based on their density: lighter particles tend to segregate towards the rotational axis, whereas thicker particles tend to go away from the axis of rotation.

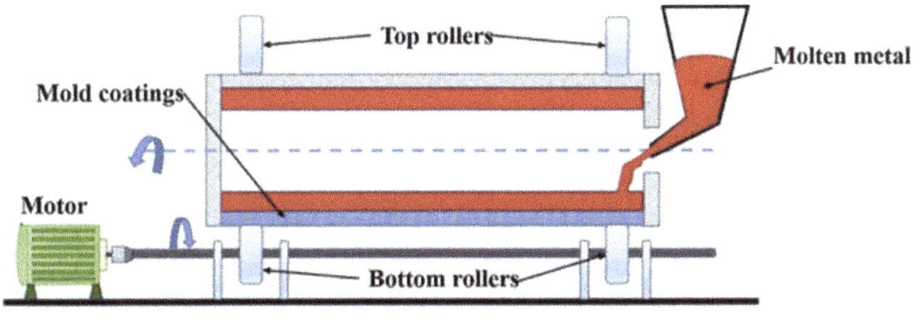

Figure 8.1. Centrifugal casting methods. Reprinted from [14], Copyright (2019), with permission from Elsevier.

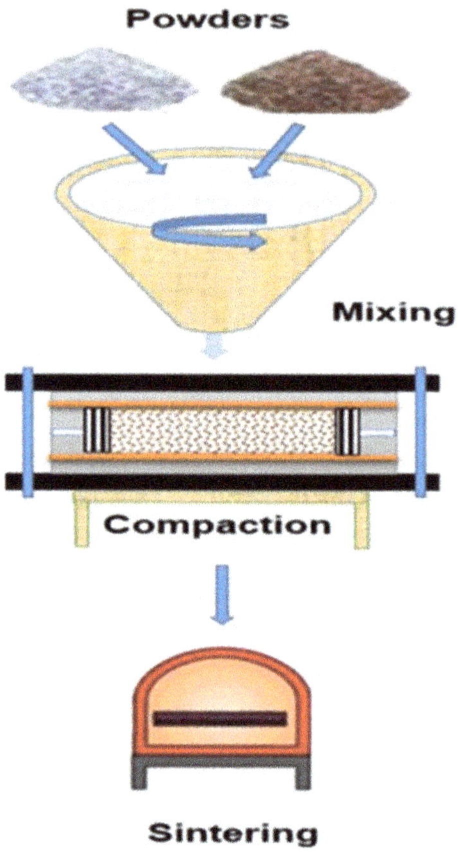

Figure 8.2. Powder metallurgy technique [16]. Reprinted from [19], Copyright (2021), with permission from Elsevier.

8.2.2.2 Powder metallurgy

Powder metallurgy is a fabrication technique commonly utilized for FGMs at micro- and nanoscales [9], employing solid materials in powder form. The critical stages of this technique include powder preparation, material processing, and forming and sintering processes, as shown in figure 8.2 [15]. In powder preparation, methods like grinding, deposition, or chemical reactions are employed to produce a high powder volume. The production rate can also be regulated to achieve the desired particle size. The working environment (temperature and pressure) influences powder sampling and distribution during the forming (piling and pressing) and sintering (powder consolidation) levels [13].

8.2.2.3 Plasma spraying

Thermal spraying processes are commonly known by various names, including plasma spraying, metal spraying, high-velocity oxygen fuel (HVOF) spraying, and arc spraying [17] to develop FG materials at micro- and nanoscales [18]. For example, a porous ceramic layering to a component with less corrosion resistance often found an

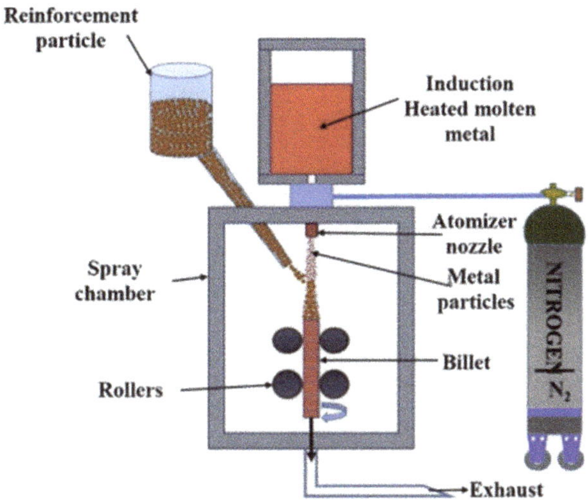

Figure 8.3. Thermal (plasma) spraying technique. Reprinted from [14], Copyright (2019), with permission from Elsevier.

application in the high-temperature areas of aero-engines [13]; high-velocity powder particles with a higher melting point are melted and propelled towards the substrate by the heat generated from the plasma spray as shown in figure 8.3. The desired layer thickness is attained through consecutive movements of the plasma spray.

8.2.2.4 Vapor deposition method

Two primary vapor deposition techniques exist: CVD and PVD, which fabricate FG materials at micro- and nanoscales [18]. While these methods exclusively provide surface coatings and are unsuitable for bulk coatings, they exhibit strong bonding properties [19]. This technique enables precise control over deposition thickness and spacing between layers, making it suitable for creating laminated grading. Vapor deposition is an effective method for applying functionally graded coatings with exceptionally fine microstructure onto surfaces of aero-engine components [13] and semiconductor applications [20]. The classification of the vapor deposition method is described in figure 8.4.

8.2.2.5 Physical vapor deposition

PVD is characterized by a process wherein the material transitions from a condense state to a vapor phase and subsequently returns to a thin condense film, as depicted in figure 8.5(a). The PVD method stands as a pivotal technique in developing thin FGNMS owing to its significant advantages. It can generate highly pure thin films with a graded structure and produce various elements, from bare metal deposition to alloys. Additionally, it is considered environmentally friendly compared to other techniques [21]. This method finds application in various fields, including aerospace and automotive industries, because it enhances mechanical properties and wear resistance by forming thin layers with graded properties [22].

Vapour deposition technique

- Physical vapour deposition (PVD)
  - Cathodic arc deposition
  - Electron beam deposition
  - Evaporative deposition
  - Close-space sublimation
  - Pulsed laser deposition
  - Sputter deposition
  - Sublimation sandwich method
- Chemical vapour deposition (CVD)
  - Atmospheric pressure
  - Low pressure
  - Ultrahigh vaccum
  - Hot filament
  - Laser assisted
  - Electron assisted
  - Direct liquid injection

Figure 8.4. Classification of vapor deposition method [20].

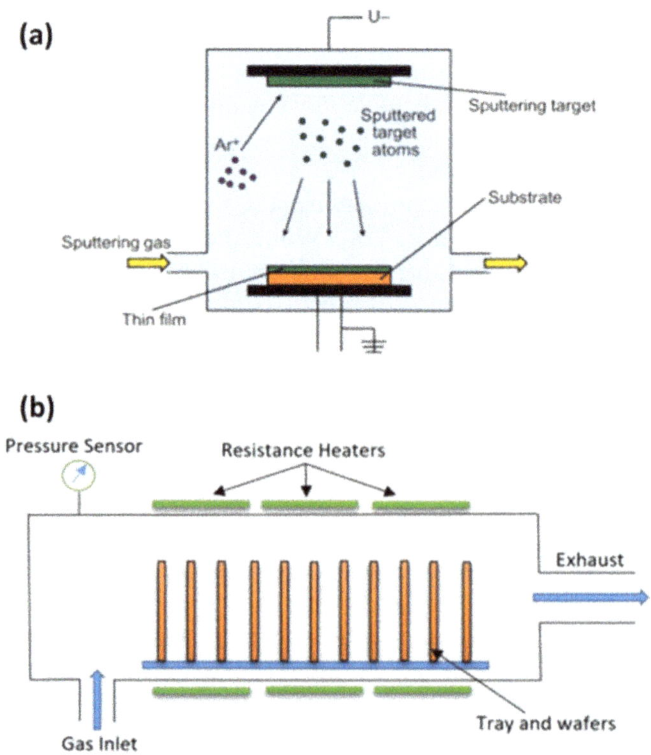

Figure 8.5. Schematic view of (a) physical vapor deposition method and (b) chemical vapor deposition method [16]. Reprinted from [16], Copyright (2021), with permission from Elsevier.

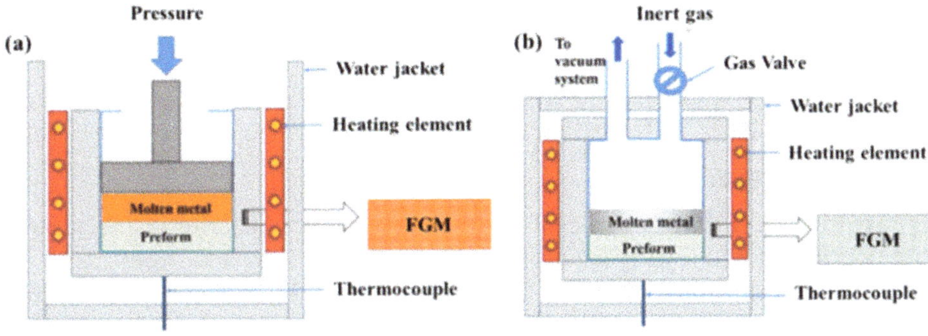

Figure 8.6. Pressure infiltration method. Reprinted from [14], Copyright (2019), with permission from Elsevier.

8.2.2.6 Chemical vapor deposition process

CVD is utilized in the manufacturing of high-quality solid materials through a vacuum deposition system, as shown in figure 8.5(b). Various CVD techniques are employed to create smooth, thin films with graded properties, including atmospheric pressure, low pressure, ultrahigh vacuum, hot filament, laser-assisted, electron-assisted, and direct liquid injection methods [23]. In traditional CVD, the substrate is exposed to one or more volatile precursors, which react with the substrate and decompose on its surface. This process involves coating the substance inside a vacuum chamber and vaporizing the coating material until it becomes vaporized [24, 25].

8.2.2.7 Lamination and infiltration methods

Infiltration is a liquid-state process used to produce FGNMS, wherein a molten matrix permeates the space among dispersed stages containing a preformed discrete phase [26]. The infiltration method using liquid metal is attained either through capillary action without pressure or by applying pressure through gaseous or mechanical means, as illustrated in figure 8.6. A pressure-infiltration technique is utilized preparing FGMs with continuous gradients. It involves using a porous mold filled with a slurry made of the desired material [14, 27]. The slurry is fed through a cavity in the mold, and after filling, the liquid part is absorbed by the mold via capillary action. Material particles that are more significant than the pores' size remain in the mold. The excess slurry is drained for hollow casting after achieving the desired thickness [28].

8.2.2.8 Solid freeform fabrication or additive manufacturing

The additive manufacturing (AM) method represents a solid freeform manufacturing technology capable of accurately fabricating FGNMS to create a pre-designed 3D object with precision [29, 30] as shown in figure 8.7. Functionally graded additive manufacturing (FGAM) methods encompass several categories, which include direct energy deposition (DED), material extrusion, material jetting, powder bed fusion (PBF), sheet lamination, and vat photopolymerization [18]. In recent years, numerous AM methods have gained popularity across various applications, including architecture, medicine, robotics, automotive, aerospace, energy, and sports. These applications demand graded properties, as they enable the processing of a diverse range of materials [31].

8.2.3 Material selection criteria

Manufacturing of FGMs at micro/nanoscale, the selection criteria for materials should consider the following aspects:

- **Material selection**: Selection of suitable materials that can be blended to achieve the desired gradient properties. This might involve choosing various compositions, phases, or microstructures [32].
- **Gradient type:** Based on the requirements of the application and the manufacturing process, the gradient that is needed is chosen, such as thickness, composition, or structure [33].
- **Application requirements:** The specific properties essential for the intended task, such as thermal stability, mechanical strength, wear resistance, or corrosion resistance are evaluated based on the application of FGNMS.
- **Cost and environment:** Based on the cost and environmental impact of both, materials and production processes are chosen. This assessment also considers material availability, energy use, and waste generation [32].
- **Design and modeling**: Utilizing suitable methodologies and modeling tools to depict the geometric and material characteristics of the FGM structure. This could entail utilizing fiber bundle spaces or multi-material tree structures to store material information [18].
- **Material distribution**: Choose a method to represent heterogeneous materials, such as a multi-material tree structure or a multi-material distribution model [18].
- **Process repeatability and reliability:** Verify the manufacturing process can produce results that are predictable, consistent, and free from flaws or variances in the gradient properties [32].
- **Adaptability for mass production:** Consider the manufacturing process's flexibility for mass production, as well as its capacity to generate significant quantities of FGM components with the required properties [32].

8.3 Functionalized MEMS and NEMS devices

8.3.1 Sensing applications

FGNMS, which modify characteristics spatially to improve functionality, have various energy and sensor systems applications [34]. They are used in thermoelectric, dielectrics, piezoelectric, and composite electrodes for solid oxide fuel cells. They offer advantages such as high capacitance, low losses, and stable properties across a broad temperature range [35]. Furthermore, functionally graded micro/nanomaterials are employed in energy absorption systems, soft robotics, and radar-absorbing structures (RAS). Impact absorption is improved in soft robots via graded mechanical characteristics. Density-graded lattices demonstrate excellent energy absorption, ideal for personal protective equipment and packaging. Moreover, radar-absorbing structures with graded designs exhibit enhanced absorption capabilities for stealth applications [18]. In manufacturing, laser-directed energy deposition (L-DED) facilitates the creation of multi-material and FGNMS parts. This technology significantly contributes to advancing the commercialization of FGNMS components by enabling the fabrication of intricate parts with diverse material compositions [36].

FGNMS are used in various sensors due to their ability to improve sensitivity, stability, and temperature dependence. Some examples of FGNMS used in sensors include [35]:

1. Thermoelectric materials
2. Dielectric materials
3. Piezoelectric materials
4. Graded optical materials
5. Graded magnetic materials

8.3.2 Thermoelectric devices

Researchers have developed a multilayer poly-SiGe deposition method to create MEMS structural films that meet material and economic constraints and can be combined with devices on the same substrate. This technique combines a CVD crystallization film with a high-growth rate Plasma-enhanced chemical vapor deposition (PECVD) bulk layer, resulting in high-quality films with outstanding electrical and mechanical properties at low temperatures ($\leqslant 450$ °C) and high deposition rates (100 nm min^{-1}) [37].

8.3.3 Optical switches

FGNMS has applications in optoelectronic devices, including optical switches, among other applications. These materials are characterized by composite structures with gradual changes in their compositions and structures. Research indicates that FGNMS can be used in optical switches for nonvolatile, all-optically modulated switching. This application integrates Ge2Sb2Te5 (GST) into graded-index multimode fibers [38]. Some examples of optical switches that utilize FGNMS include:

- **MEMS optical switches:** MEMS-based optical switches perform on the basis of micromirrors to send the required ouput optical signals. Additionally, these switches can be functionalized with different materials to enhance their performance [39].
- **NEMS optical switches:** NEMS-based optical switches work at nanoscales, presenting smaller footprints and faster response times than MEMS devices. Functionalizing NEMS devices can improve their sensitivity and efficiency in optical switching applications.
- **Commercial applications:** MEMS- and NEMS-based optical switches have been deployed in large numbers for commercial applications, showcasing their reliability and scalability in real-world settings [40].

8.3.4 Resonators

Incorporating FGNMS into resonators improves their performance and characteristics. Studies have explored using FGNMS in resonators to improve wave-propagation properties, create single-frequency piezoelectric resonators, and analyze elastic materials. These applications are intended to benefit from the unique characteristics of FGNMS in order to maximize resonator performance. Applying FGNMS instead of homogeneous materials in resonator design can improve mechanical properties and increase effectiveness [41–43].

- **Hybrid lattice metamaterials:** Structures composed of functionally graded metamaterials have been utilized to improve the mechanical properties of homogeneous materials, highlighting the versatility of FGNMS in resonator applications [41].
- **Piezoelectric resonators:** Resonators with single-frequency capability have been designed using functionally graded piezoelectric materials to improve the mechanical performances [42].
- **Elastic materials:** Quartz crystal plate thickness-shear mode resonators have been used to characterize elastic materials graded functionally, showing the ability to measure material property fluctuations within FGNMS [43].

8.3.5 Pressure sensors

FGNMS finds various applications, notably in pressure sensors, due to their continuously graded properties, enabling spatially varying microstructures with nonuniform compositions. They are used in many industries, including biomedicine, dental implants, energy, and sensors [20, 35]. In the context of pressure sensors, functionally graded dielectric films are employed in capacitors exhibiting low-temperature coefficients. These films utilize perovskite-type alkaline earth titanates, valued for their high dielectric constant and minimal temperature dependence [35]. Similarly, pressure sensors incorporating FGNMS include:

- **Flexible piezoresistive sensors:** Using a graded nest-like architecture, these sensors achieve wide-ranging pressure measurements while improving sensitivity and responsiveness [44].
- **Ultra-sensitive resistive pressure sensor:** The sensor relies on an elastic, microstructured conducting polymer thin film, allowing the detection of pressures below 1 Pa with exceptional sensitivity and cycling stability [45].
- **Functionally graded dielectric films:** Perovskite-type alkaline earth titanates used in capacitors with low-temperature coefficients, providing a high dielectric constant and minimal temperature dependence, thus enhancing performance [35].

8.3.6 Accelerometers

The design of electrostatically actuated microswitches with FGNMS involves functionally graded microbeams bonded with piezoelectric layers under electric force, considering both static and dynamic aspects [46]. Moreover, the utilization of

FGM structures is rising in engineering because of their exceptional mechanical and material properties. The vibration analysis of functionally graded rectangular plates with cut-outs offers valuable insights into the behavior of FGNMS structures under various conditions, providing benchmark data for further research in this domain [47]. Examples of accelerometer sensors incorporating FGNMS include:

- **Piezoelectric sensors:** FGNMS are employed in piezoelectric sensors to optimize output voltage by reducing the stiffness of piezoelectric materials, thus enhancing sensor performance [48].
- **Vibration control systems:** FGNMS are used in vibration control systems to demonstrate its efficacy in this field when active vibration control for FGM plates is utilized via piezoelectric sensor/actuator patches [49].

8.3.7 Gyroscopes

These materials demonstrate a progressive alteration in their properties, enabling the creation of optimized combinations that improve the functionality of gyroscopes. Studies have investigated utilizing material distributions in FGNMS to adjust gyroscope and energy harvesting performance [50]. FGMs help structures transition smoothly between different material properties, achieving optimal gyroscope performance [51]. Furthermore, FGNMS has been found to be applicable in aerospace settings, such as aircraft structures and gyroscopes. The gradual alteration of material properties in FGMs offers advantages for improving the performance and efficiency of these components [52, 53].

8.3.8 Actuation systems

The utilization of FGNMS holds the potential to enhance the performance of the actuation system [34]. Actuators, which transform energy into mechanical motion, find application across diverse sectors, including robotics, aerospace, and automotive industries. FGNMS in actuation systems offers various potential applications [18], such as:

- **Improved thermal management:** FGNMS can be engineered to possess superior thermal conductivity and heat dissipation characteristics. These attributes aid in regulating the actuator's temperature and enhancing its overall performance [54].
- **Enhanced mechanical properties:** Through the grading of material composition, FGNMS can be customized to exhibit enhanced mechanical properties like strength, stiffness, and toughness. This customization holds the potential to result in actuators that are more efficient and resilient over time.
- **Optimized energy conversion:** FGNMS can undergo design modifications to enhance energy conversion efficiency, thereby contributing to the development of more energy-efficient actuators.
- **Reduced stress concentrations:** The gradual alteration of material properties can assist in reducing stress concentrations within the actuator, ultimately resulting in a prolonged lifespan and increased reliability.

8.3.9 Microfluidic devices

FGNMS finds wide applications across different domains, particularly in microfluidic devices. Engineered to replicate the inherent gradients observed in biological tissues, FGNMS are crucial in altering mechanical attributes and cellular communication. Within microfluidic systems, FGNMS enables the generation of gradients encompassing substrate stiffness, oxygen levels, and chemical compositions akin to those witnessed in natural physiology but on a miniature scale. Such gradient formations are invaluable for conducting rapid, high-throughput assessments of chemical impacts on cellular behavior and evaluating treatment effectiveness across diverse domains, notably in cancer research [55]. FGNMS integrated into microfluidic devices offers versatile applications across multiple sectors, including biology, aerospace, and energy.

- **Biomaterials:** FGNMS allows the study of tissue engineering and cell behavior using model systems. They can replicate the inherent gradients observed in biological tissues, influencing mechanical characteristics and physiological activity. This capability enables high-throughput analysis of the effects of chemical concentration on cells and the evaluation of treatment efficiency in various sectors, including cancer research [55].
- **Aerospace:** FGNMS are used to provide high temperature, thermal shock, thermal fatigue, and corrosion resistance to aviation engine and spacecraft components, thereby enhancing their resistance to heat. Subsequently, they have outstanding heat insulation qualities, heat resistance, and resistance to thermal degradation, withstanding temperatures above 2000 K [56].
- **Energy conservation:** FGNMS have potential applications in power generation systems, particularly thermoelectric energy conversion materials. They assist in decreasing heat stress and improving the effectiveness of energy conversion systems [56].

8.4 Structural health monitoring

Due to their unique properties and capabilities, FGNMs offer promising applications in structural health monitoring (SHM). Here are some potential applications:

- Sensing: FGNMs designed with embedded sensors capable of detecting various parameters such as strain, stress, temperature, and vibration. These sensors can provide real-time data on the material's structural integrity and alert it to any potential defects or damage [35].
- Damage detection: By monitoring changes in the properties of the FGNSMs, such as electrical conductivity or acoustic impedance, it is possible to detect and locate damage, such as cracks or delamination, in structures. The spatial distribution of properties in FGNSMs allows for more precise damage localization compared to homogeneous materials [57].
- Self-healing materials: FGNMs are engineered to possess self-healing capabilities at the nano- or microscale. By integrating microcapsules containing healing agents or incorporating nanoparticles with catalytic properties,

FGNMs can autonomously repair damage, thus enhancing the durability and longevity of structures [58].
- Adaptive structures: FGNMs exhibit adaptive behavior in response to changing environmental conditions or applied loads. By adjusting their properties based on external stimuli, such as temperature, moisture, or mechanical stress, FGNMs can optimize their performance and mitigate potential damage [59].
- Energy harvesting: FGNMs are utilized to harvest energy from ambient vibrations or mechanical deformations experienced by structures during operation. Piezoelectric or triboelectric materials integrated into FGNMs can convert mechanical energy into electrical energy, which can be used to power sensors or other monitoring devices [60].
- Reduced weight and improved performance: By tailoring the composition and structure of FGNMs, it is possible to achieve lightweight materials with enhanced mechanical properties, such as strength, stiffness, and toughness. These materials can be used to construct lighter, more durable structures without compromising performance [61].
- Real-time monitoring and feedback: Integrating FGNMs with wireless communication systems enables real-time monitoring of structural health conditions. Data collected from sensors embedded in FGNMs can be transmitted wirelessly to a central monitoring station, providing timely feedback on the material's structural integrity and facilitating proactive maintenance and repair strategies [62].

8.5 Challenges and future perspectives

Several issues need to be addressed and resolved when applying FGMs in MEMS and NEMS. Some of them are listed below in the subsequent section [63].

- The complete analysis of gradient components, preparation, performance, and parameters must be established.
- Expertise is required while pouring and stacking materials in layers.
- More research on microscopic structure and preparation conditions is needed to precisely predict graded material's physical properties.
- A complete investigation of the improved preparation process mechanism for the mass production of FGM is required.
- Manufacturing costs are extremely expensive.

8.5.1 Integration and compatibility issues

Flaws in the FG materials are mostly caused by debonding and cracking due to rapid changes in material characteristics under high thermal or mechanical stress conditions [57] and poor wetting between the matrix and reinforcement [64].

8.5.2 Reliability and durability

FGNMS has several issues, especially in terms of dependability and durability. Accurate control of material composition and microstructure gradients is challenging resulting in variations that may undermine reliability [65]. Stress concentrations can arise at the interfaces between the various graded layers, causing interfacial delamination and reduced durability [66]. Fabricating FGNMS with controlled gradients usually needs sophisticated processing methods, which raise production costs and create potential faults [67]. The complexity of graded structures makes it challenging to anticipate the mechanical behavior and performance of FGNMS under varied loading conditions. Scaling up the fabrication of FGNMS for industrial applications while maintaining reliability and durability presents significant challenges, especially when adding graded structures into complicated systems [67].

8.6 Emerging trends and opportunities

FGNMS may be modified and quickly prototyped for a wide range of applications, including aviation components, medical implants, and microfluidic devices, using additive manufacturing techniques with complex shapes and precise control over material gradients [18]. Bioinspired materials are being developed for tissue engineering, lightweight construction, and energy harvesting applications. Also, using natural structures as inspiration, hierarchical FGNMS are designed with higher mechanical strength, durability, and biomimetic properties [68]. Furthermore, to accelerate the development of new materials and optimize material processing parameters to improve dependability, durability, and performance, machine learning and computational modeling techniques are employed to improve the design and manufacture of FGNMS with customized properties [69]. Self-healing FGNMS are being developed to improve the reliability and durability of structural materials, electronic components, and biomedical implants, including self-healing mechanisms to automatically repair damage and extend the operating life of materials and devices [70]. Sustainable FGNMS development for renewable energy, pollution cleaning, and environmentally friendly consumer products utilizing green synthesis procedures and eco-friendly materials has recently been trending [71]. FGNMs have many potential uses in microthrusters and micropropulsion systems, which are critical for small satellite propulsion, microrobotics, and other miniaturized propulsion technologies. Microthrusters generate a large amount of heat while operating. FGNMs can be created with graded thermal conductivity, enabling efficient heat dissipation and thermal control within the microthruster system. FGNMs can prevent overheating by managing thermal conductivity distribution at the nano- and microscale, ensuring the dependability and durability of the propulsion system [72]. FGNMs can be programmed to regulate the release and distribution of propellant gases in microthrusters. FGNSMs may precisely control the flow of propellant gases by including nanoscale pores or channels into the material structure, allowing for fine thrust modulation and maneuverability in micropropulsion systems [73]. FGNMs can use catalytic nanoparticles or nanoscale features on their surfaces to improve propellant combustion efficiency. These catalytic surfaces enable quick and complete combustion of the propellant gas, resulting in greater thrust output and improved propulsion performance in microthrusters [74]. Microthrusters and micropropulsion devices work under challenging situations with high temperatures, mechanical vibrations, and radiation exposure. FGNSMs can tolerate these harsh circumstances by optimizing mechanical qualities such as strength, stiffness, and fatigue resistance. This ensures the propulsion system remains structurally sound and durable throughout its operational life [75]. FGNSMs allow for the miniaturization and integration of propulsion components into tiny microsystems. FGNSMs, which fabricate propulsion components with nano- and microscale precision, enable the construction of lightweight and compact microthrusters that can be seamlessly incorporated into small satellites, unmanned aerial vehicles (UAVs), and other microrobotic platforms [31]. FGNSMs can use energy harvesting methods, such as piezoelectric or thermoelectric materials, to generate electricity from mechanical or thermal energy generated during propulsion operations. This self-powering capability decreases the dependency on external power sources and increases the autonomy of micropropulsion devices [76]. FGNSMs can be used in microthrusters to provide dynamic control and accurate maneuverability for microspacecraft and microrobots. The thrust vector and orientation of the propulsion system can be adjusted by changing the characteristics of FGNSMs in reaction to external stimuli, such as electric or magnetic fields. This allows for nimble and responsive motion control in limited places [18]. Overall, the use of functionally graded nano- and microstructures in microthrusters and micropropulsion systems provides significant benefits in terms of thermal management, propellant control, structural integrity, miniaturization, energy harvesting, and maneuverability, making them critical for next-generation microsatellite propulsion and microrobotics applications.

8.7 Conclusion

FGNMs have emerged as essential components in MEMS and NEMS, providing unrivaled capability and efficiency across a wide range of applications. As research in materials science, fabrication processes, and system integration advances, this field has enormous potential for further breakthroughs and innovations. By addressing the present issues and utilizing accessible technologies, the future appears optimistic for the continuous advancement and broad deployment of functionally graded nano- and microstructures in MEMS and NEMS.

References

[1] Sun Y, Cheng J, Wang Z, Yu Y, Tian L and Lu J 2019 Analytical approximate solution for nonlinear behavior of cantilever FGM MEMS beam with thermal and size dependency *Math. Probl. Eng.* **2019** 1–10

[2] Taati E and Sina N 2018 Static pull-in analysis of electrostatically actuated functionally graded micro-beams based on the modified strain gradient theory *Int. J. Appl. Mech.* **10** 1850031

[3] Haghshenas Gorgani H, Mahdavi Adeli M and Hosseini M 2019 Pull-in behavior of functionally graded micro/nano-beams for MEMS and NEMS switches *Microsyst. Technol.* **25** 3165–73

[4] Sedighi H M, Daneshmand F and Abadyan M 2015 Dynamic instability analysis of electrostatic functionally graded doubly-clamped nano-actuators *Compos. Struct.* **124** 55–64

[5] Yang W D, Yang F P and Wang X 2016 Coupling influences of nonlocal stress and strain gradients on dynamic pull-in of functionally graded nanotubes reinforced nano-actuator with damping effects *Sens. Actuators,* A **248** 10–21

[6] Sadeghzadeh S and Mahinzare M 2020 Free vibration analysis of a spinning smart piezoelectrically actuated heterogeneous nanoscale shell with nonlocal strain gradient theory *J. Nano Res.* **64** 1–19

[7] Shojaeefard M H, Saeidi Googarchin H, Mahinzare M and Adibi M 2018 Vibration and buckling analysis of a rotary functionally graded piezomagnetic nanoshell embedded in viscoelastic media *J. Intell. Mater. Syst. Struct.* **29** 2344–61

[8] Mahinzare M, Jannat Alipour M, Sadatsakkak S A and Ghadiri M 2019 A nonlocal strain gradient theory for dynamic modeling of a rotary thermo piezo electrically actuated nano FG circular plate *Mech. Syst. Signal Process.* **115** 323–37

[9] Kieback B, Neubrand A and Riedel H 2003 Processing techniques for functionally graded materials *Mater. Sci. Eng.* A **362** 81–106

[10] Ghayesh M H and Farajpour A 2019 A review on the mechanics of functionally graded nanoscale and microscale structures *Int. J. Eng. Sci.* **137** 8–36

[11] Zhang N, Khan T, Guo H, Shi S, Zhong W and Zhang W 2019 Functionally graded materials: an overview of stability, buckling, and free vibration analysis *Adv. Mater. Sci. Eng.* **2019** 1–18

[12] Rajan T P D, Pillai R M and Pai B C 2008 Centrifugal casting of functionally graded aluminium matrix composite components *Int. J. Cast Met. Res.* **21** 214–8

[13] Popoola P A I, Farotade G, Fatoba O S and Popoola O 2016 Laser engineering net shaping method in the area of development of functionally graded materials (FGMs) for aero engine applications—a review *Fiber Laser* (Rijeka: InTech)

[14] Ramanathan A, Krishnan P K and Muraliraja R 2019 A review on the production of metal matrix composites through stir casting—furnace design, properties, challenges, and research opportunities *J. Manuf. Processes* **42** 213–45

[15] Parihar R S, Setti S G and Sahu R K 2018 Recent advances in the manufacturing processes of functionally graded materials: a review *Sci. Eng. Compos. Mater.* **25** 309–36

[16] Sam M, Jojith R and Radhika N 2021 Progression in manufacturing of functionally graded materials and impact of thermal treatment—a critical review *J. Manuf. Processes* **68** 1339–77

[17] Ganesh V, Dorow-Gerspach D, Heuer S, Matejicek J, Vilemova M, Bram M *et al* 2021 Manufacturing of W-steel joint using plasma sprayed graded W/steel-interlayer with current assisted diffusion bonding *Fusion Eng. Des.* **172** 112896

[18] Li Y *et al* 2020 A review on functionally graded materials and structures via additive manufacturing: from multi-scale design to versatile functional properties *Adv. Mater. Technol.* **5**

[19] Sarathchandra D T, Kanmani Subbu S and Venkaiah N 2018 Functionally graded materials and processing techniques: an art of review *Mater. Today Proc.* **5** 21328–34

[20] Saleh B, Jiang J, Fathi R, Al-hababi T, Xu Q, Wang L *et al* 2020 30 years of functionally graded materials: an overview of manufacturing methods, applications and future challenges *Composites* B **201** 108376

[21] Rajak D K, Wagh P H, Menezes P L, Chaudhary A and Kumar R 2020 Critical overview of coatings technology for metal matrix composites *J. Bio-Tribo-Corros.* **6** 12

[22] Shahidi S, Moazzenchi B and Ghoranneviss M 2015 A review-application of physical vapor deposition (PVD) and related methods in the textile industry *Eur. Phys. J. Appl. Phys.* **71** 31302
[23] Oluwatosin Abegunde O, Titilayo Akinlabi E, Philip Oladijo O, Akinlabi S and Uchenna Ude A 2019 Overview of thin film deposition techniques *AIMS Mater. Sci.* **6** 174–99
[24] Reina A, Jia X, Ho J, Nezich D, Son H, Bulovic V, Dresselhaus M S and Kong J 2009 Large area, few-layer graphene films on arbitrary substrates by chemical vapor deposition *Nano Lett.* **9** 30–5
[25] Choy K 2003 Chemical vapour deposition of coatings *Prog. Mater Sci.* **48** 57–170
[26] Sree Manu K M, Resmi V G, Brahmakumar M, Narayanasamy P, Rajan T P D, Pavithran C *et al* 2012 Squeeze infiltration processing of functionally graded aluminum–SiC metal ceramic composites *Trans. Indian Inst. Met.* **65** 747–51
[27] Rajan T P D and Pai B C 2014 Developments in processing of functionally gradient metals and metal–ceramic composites: a review *Acta Metall. Sin. Engl. Lett.* **27** 825–38
[28] Janković Ilić D, Fiscina J, González-Oliver C J R and Mücklich F 2005 Properties of Cu–W functionally graded materials produced by segregation and infiltration *Mater. Sci. Forum* 492–3 123–8
[29] Alkunte S, Fidan I, Naikwadi V, Gudavasov S, Ali M A, Mahmudov M *et al* 2024 Advancements and challenges in additively manufactured functionally graded materials: a comprehensive review *J. Manuf. Mater. Process.* **8** 23
[30] Loh G H, Pei E, Harrison D and Monzón M D 2018 An overview of functionally graded additive manufacturing *Addit. Manuf.* **23** 34–44
[31] Yan L, Chen Y and Liou F 2020 Additive manufacturing of functionally graded metallic materials using laser metal deposition *Addit. Manuf.* **31** 100901
[32] El-Galy I M, Saleh B I and Ahmed M H 2019 Functionally graded materials classifications and development trends from industrial point of view *SN Appl. Sci.* **1** 1378
[33] Ghanavati R and Naffakh-Moosavy H 2021 Additive manufacturing of functionally graded metallic materials: a review of experimental and numerical studies *J. Mater. Res. Technol.* **13** 1628–64
[34] Wang K, Hu T, Zhao Y, Ren W and Liu J 2022 Design of a double-layer electrothermal MEMS safety and arming device with a bistable mechanism *Micromachines* **13** 1076
[35] Müller E, Drašar Č, Schilz J and Kaysser W 2003 Functionally graded materials for sensor and energy applications *Mater. Sci. Eng.* A **362** 17–39
[36] Wasmer K, Wüst M, Cui D, Masinelli G, Pandiyan V and Shevchik S 2023 Monitoring of functionally graded material during laser directed energy deposition by acoustic emission and optical emission spectroscopy using artificial intelligence *Virtual Phys. Prototyping* **18**
[37] Witvrouw A and Mehta A 2005 The use of functionally graded poly-SiGe layers for MEMS applications *Mater. Sci. Forum* 492–3 255–60
[38] Zhang Y, Liu S, Chen J, Cheng S, Jin W, Zhang Y *et al* 2022 All-optically modulated nonvolatile optical switching based on a graded-index multimode fiber *Opt. Express* **30** 36691
[39] Yeow T-W, Law K L E and Goldenberg A 2001 MEMS optical switches *IEEE Commun. Mag.* **39** 158–63
[40] Berman D and Krim J 2013 Surface science, MEMS and NEMS: progress and opportunities for surface science research performed on, or by, microdevices *Prog. Surf. Sci.* **88** 171–211

[41] Jafari H, Sepehri S, Yazdi M R H, Mashhadi M M and Fakhrabadi M M S 2020 Hybrid lattice metamaterials with auxiliary resonators made of functionally graded materials *Acta Mech.* **231** 4835–49

[42] Rubio W M, Silva E C N and Paulino G H 2009 Topology optimized design of functionally graded piezoelectric resonators with specified resonance frequencies *Mater. Sci. Forum* 631–2 305–10

[43] Chen Y, Wang J, Du J and Yang J 2013 Characterization of functionally graded elastic materials using a thickness-shear mode quartz resonator *Philos. Mag. Lett.* **93** 362–70

[44] Seesaard T and Wongchoosuk C 2023 Flexible and stretchable pressure sensors: from basic principles to state-of-the-art applications *Micromachines* **14** 1638

[45] Pan L, Chortos A, Yu G, Wang Y, Isaacson S, Allen R *et al* 2014 An ultra-sensitive resistive pressure sensor based on hollow-sphere microstructure induced elasticity in conducting polymer film *Nat. Commun.* **5** 3002

[46] Hosseinzadeh A and Ahmadian M T 2010 Application of piezoelectric and functionally graded materials in designing electrostatically actuated micro switches *Volume 4: 12th Int. Conf. on Advanced Vehicle and Tire Technologies; 4th Int. Conf. on Micro- and Nanosystems* (ASMEDC) pp 613–20

[47] Huang W, Xue K and Li Q 2021 Three-dimensional solution for the vibration analysis of functionally graded rectangular plate with/without cutouts subject to general boundary conditions *Materials* **14** 7088

[48] Chandrasekaran S 2021 Functionally graded materials for marine risers *Design of Marine Risers with Functionally Graded Materials* (Amsterdam: Elsevier) pp 25–58

[49] Kargarnovin M H, Najafizadeh M M and Viliani N S 2007 Vibration control of a functionally graded material plate patched with piezoelectric actuators and sensors under a constant electric charge *Smart Mater. Struct.* **16** 1252–9

[50] Serrano M, Larkin K and Abdelkefi A 2023 Performance analysis of functionally graded multifunctional piezoelectric energy harvesting microgyroscopes *Int. J. Non Linear Mech.* **157** 104521

[51] Dastjerdi S, Akgöz B and Civalek Ö 2020 On the effect of viscoelasticity on behavior of gyroscopes *Int. J. Eng. Sci.* **149** 103236

[52] Miteva A and Bouzekova-Penkova A 2021 Module for wireless communication in aerospace vehicles *Aerosp. Res. Bulg.* **33** 195–209

[53] Sedebo G T, Shatalov M Y, Joubert S V and Shafi A A 2022 The dynamics of a three-dimensional tuning functionally graded plate gyroscope *Mech. Solids* **57** 1577–89

[54] Ahankari S S and Kar K K 2017 Functionally graded composites: processing and applications *Composite Materials* (Berlin: Springer) pp 119–68

[55] Lowen J M and Leach J K 2020 Functionally graded biomaterials for use as model systems and replacement tissues *Adv. Funct. Mater.* **30**

[56] Li W and Han B 2018 Research and application of functionally gradient materials *IOP Conf. Ser.: Mater. Sci. Eng.* **394** 022065

[57] Sabri N, Aljunid S A, Salim M S and Fouad S 2015 Fiber optic sensors: short review and applications *Springer Series Materials Science* **vol 204** (Berlin: Springer) pp 299–311

[58] Yang Y and Urban M W 2013 Self-healing polymeric materials *Chem. Soc. Rev.* **42** 7446

[59] Bhavar V, Kattire P, Thakare S, Patil S and Singh R 2017 A review on functionally gradient materials (FGMs) and their applications *IOP Conf. Ser.: Mater. Sci. Eng.* **229** 012021

[60] Covaci C and Gontean A 2020 Piezoelectric energy harvesting solutions: a review *Sensors* **20** 3512

[61] Shi H, Zhou P, Li J, Liu C and Wang L 2021 Functional gradient metallic biomaterials: techniques, current scenery, and future prospects in the biomedical field *Front. Bioeng. Biotechnol.* **8**

[62] Aqeel-ur-Rehman , Abbasi A Z, Islam N and Shaikh Z A 2014 A review of wireless sensors and networks' applications in agriculture *Comput. Stand. Interfaces* **36** 263–70

[63] Pasha A and Rajaprakash B M 2022 Functionally graded materials (FGM) fabrication and its potential challenges and applications *Mater. Today Proc.* **52** 413–8

[64] Panchal Y and Ponappa K 2022 Functionally graded materials: a review of computational materials science algorithms, production techniques, and their biomedical applications *Proc. Inst. Mech. Eng. Part C J. Mechan. Eng. Sci.* **236** 10969–86

[65] Udupa G, Rao S S and Gangadharan K V 2014 Functionally graded composite materials: an overview *Procedia Mater. Sci.* **5** 1291–9

[66] Pasha A and Rajaprakash B M 2022 Functionally graded materials (FGM) fabrication and its potential challenges & applications *Mater. Today Proc.* **52** 413–8

[67] Naebe M and Shirvanimoghaddam K 2016 Functionally graded materials: a review of fabrication and properties *Appl. Mater. Today* **5** 223–45

[68] Wei J, Pan F, Ping H, Yang K, Wang Y, Wang Q *et al* 2023 Bioinspired additive manufacturing of hierarchical materials: from biostructures to functions *Research* **6**

[69] Mueller T, Kusne A G and Ramprasad R 2016 Machine learning in materials science: recent progress and emerging applications *Rev. Comput. Chem.* **29** 186–273

[70] White S R, Sottos N R, Geubelle P H, Moore J S, Kessler M R, Sriram S R *et al* 2001 Autonomic healing of polymer composites *Nature* **409** 794

[71] Kumar Gupta G, De S, Franco A, Balu A and Luque R 2015 Sustainable biomaterials: current trends, challenges and applications *Molecules* **21** 48

[72] Sharma R, Jadon V K and Singh B 2015 A review on the finite element methods for heat conduction in functionally graded materials *J. Inst. Eng. India Ser.* C **96** 73–81

[73] Silva M A C, Guerrieri D C, Cervone A and Gill E 2018 A review of MEMS micro-propulsion technologies for CubeSats and PocketQubes *Acta Astronaut.* **143** 234–43

[74] Puchades I, Hobosyan M, Fuller L F, Liu F, Thakur S, Martirosyan K S *et al* 2014 MEMS microthrusters with nanoenergetic solid propellants *14th IEEE Int. Conf. on Nanotechnology* (Piscataway, NJ: IEEE) pp 83–6

[75] Alibeigloo A 2010 Thermoelasticity analysis of functionally graded beam with integrated surface piezoelectric layers *Compos. Struct.* **92** 1535–43

[76] Lu F, Lee H P and Lim S P 2004 Modeling and analysis of micro piezoelectric power generators for micro-electromechanical-systems applications *Smart Mater. Struct.* **13** 57–63

Chapter 9

Application of functionally graded nano- and microstructures in biomedical devices

Erukala Kalyan Kumar, Ashish Kumar Meher, Vikash Kumar and Subrata Kumar Panda

Functionally graded materials (FGMs) offer unique characteristics resulting from their precisely controlled variation in composition and structure at the nano- and microscales. The incorporation of these graded structures into biomedical devices present unprecedented opportunities for tailoring mechanical, biological, and physicochemical properties. This chapter explores the dynamic field of functionally graded (FG) nano- and microstructures in biomedical devices, examining their applications and impact on healthcare systems. It provides comprehensive insights into the diverse range of biomedical applications, including drug delivery systems, implantable devices, biosensors, and diagnostic tools. Through a critical examination of existing research and real-world case studies, this chapter is an invaluable resource for researchers, engineers, and healthcare professionals keen on harnessing the transformative capabilities of functionally graded nano- and microstructures (FGNMs) in advancing the landscape of biomedical devices.

9.1 Introduction

The evolution of manufacturing techniques has prompted a shift in material sciences from traditional metals to advanced composites and smart materials. Despite the promising attributes of these advancements, their applications are often constrained by certain limitations. Addressing these challenges, a novel category of composite materials known as functionally graded materials (FGMs) has garnered substantial attention within the scientific community due to their inherent mechanical performance. Extensive work in the literature highlights the processing and modeling of FGMs, rendering them suitable for a wide range of industries such as biomedical, automotive, and aerospace. While composite materials exhibit excellent properties, a

doi:10.1088/978-0-7503-6024-1ch9

notable drawback lies in their susceptibility to delamination failure, especially under high-temperature conditions, where fibers separate from the matrix material due to distinct expansion properties. To overcome this limitation, researchers introduced FGMs in the past, heralding a new class of composite materials also known as advanced engineering materials. FGMs demonstrate resilience in harsh environmental conditions without compromising the properties of constituent materials, thereby minimizing failures during service. The distinguishing feature of FGMs is their compositional gradient, transitioning from one material to another.

FGMs offer a pioneering method for achieving unique properties and functions that conventional homogeneous materials cannot achieve. Unlike conventional materials, where the composition or structure remains uniform throughout the volume, FGMs exhibit gradual changes in features from one layer to another, resulting in progressive variations in macroscopic and nanoproperties. Functionally graded nano- and microstructures (FGNMs) in biomedical devices have shown significant promise in various applications. These structures, which exhibit gradual variations in composition, properties, or morphology, offer unique advantages for improving the performance and biocompatibility of biomedical devices. The integration of bioinspired composites in structural innovation offers a unique perspective for creating new lightweight structures with enhanced mechanical properties. Among the noteworthy considerations in this context, porous structures stand out as a crucial topic for discussion.

Modern medicine is in dire need of advanced materials and versatile methods that can address diverse objectives. The biomedical field, focusing on technological advancements to improve the quality of human life, places immense importance on various parameters. The effectiveness of biomedical devices is heavily reliant on the choice of materials. A wide array of materials, such as metals and alloys, ceramics, composites, and polymers, find application in this field. Furthermore, the design of implants plays a pivotal role, with numerous designs aiming to emulate the functionalities of natural organs. The success of biomedical endeavors hinges on the intricate interplay of these material choices and design innovations [1]. Here are some applications of FGNMs shown in figure 9.1.

9.2 Applications of FGMs in biomedical devices

FGMs exhibit a gradient in composition and properties, making them versatile for a range of biomedical applications as shown in figure 9.2. In orthopedic implants, FGMs contribute to joint replacements, spinal implants, and fracture fixation devices, optimizing mechanical properties to reduce stress shielding and improve stability. Dental applications involve FGMs in root canal posts, implant abutments, and prosthetic crowns, enhancing the longevity and performance of restorations. FGMs also play a crucial role in drug delivery systems, influencing drug release kinetics in stents and implantable drug pumps. In tissue engineering, scaffolds with FGMs support cell growth and differentiation, facilitating tissue regeneration. Moreover, FGMs find application in wound healing, diagnostic tools, surgical instruments, and biomedical sensors, showcasing their adaptability in various clinical scenarios.

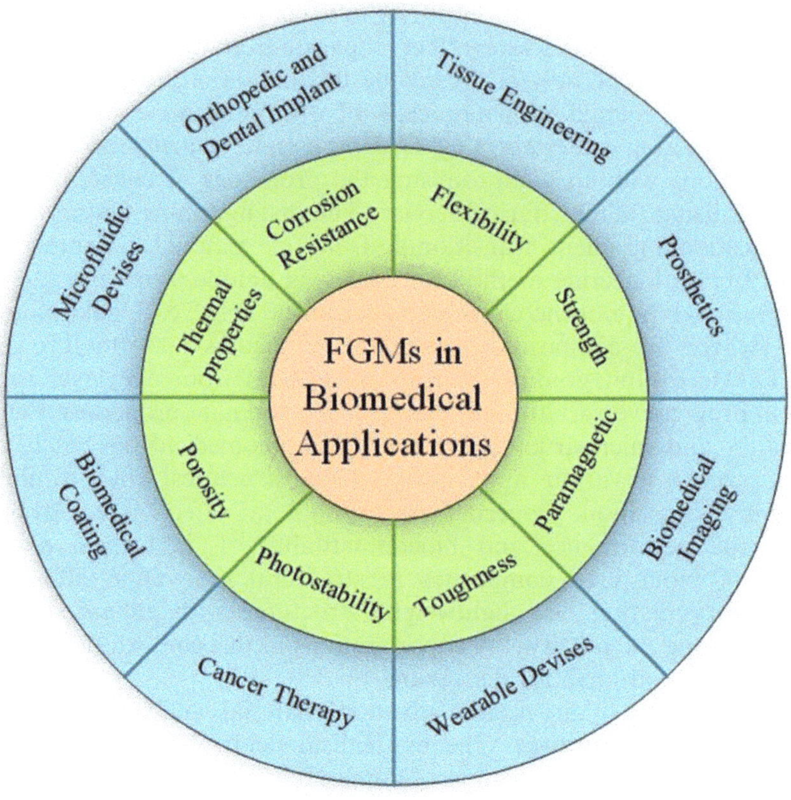

Figure 9.1. Biomedical applications and their corresponding properties of FGNMs.

This customizable and biomimetic approach holds promise for revolutionizing clinical treatments, offering tailored solutions that closely align with the complex requirements of biological systems. The adaptability and customizable nature of FGMs make them valuable in clinical applications, allowing for the development of implants and devices that closely match the biological and mechanical properties of the human body. The following sections gives an in detailed explanation of FGM applications.

9.2.1 Implantable devices

9.2.1.1 Orthopedic implants

FGNMs can be used in orthopedic implants, such as bone plates and screws, with gradient structures tailored to mimic the mechanical properties of natural bone. This helps reduce stress shielding and promote better osseointegration. With gradually changing compositions and properties, these structures can be tailored to meet the specific demands of orthopedic implants, such as compatibility with bone density and strength, structural integrity, bioactivity, corrosion resistance, and stress shielding effect [3]. Nanotechnology has also played a significant role in enhancing the properties of orthopedic implants. Nanoparticles, when incorporated into implant surfaces, can mimic

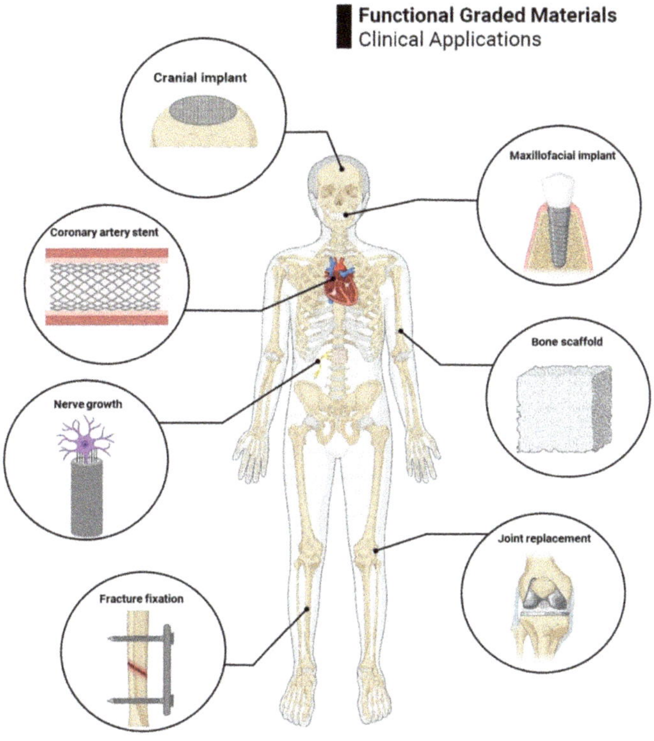

Figure 9.2. Applications of FGMs in the clinical field. Reprinted from [2], Copyright (2023), with permission from Elsevier.

the features of real tissues, improve wettability, topography, chemistry, and energy, and enhance antibacterial/antimicrobial activity, cell attachment, propagation, and differentiation [4]. Furthermore, incorporating bifunctional elements into nano/microstructure coatings on titanium and its alloys has enhanced bone repair/regeneration and bacterial resistance of orthopedic implants [5]. FGMs are extensively employed in orthopedic implants, particularly in joint replacements. In hip and knee replacements. In fracture fixation, FGMs are utilized in the development of bone plates and screws. These implants are designed with a gradual transition in mechanical properties, ensuring a smoother load transfer from the implant to the surrounding bone. This helps in reducing stress concentrations and minimizing the risk of complications during the healing process. FGMs are tailored to mimic the mechanical properties of natural bone, mitigating issues like stress shielding and improving the long-term stability of the implants. In spinal implants, the use of FGMs addresses the unique mechanical demands, providing enhanced fusion and reducing complications.

9.2.1.2 Dental implants

FGNMs are applied in dental implants to improve their mechanical and biological properties. Gradient structures in dental implants can enhance the bonding between the implant and surrounding tissues, improving stability and longevity. These structures consist of two or more constituent phases with continuous changes in microstructure, allowing for adjustable through-thickness properties [9]. Graded porous titanium scaffolds with different porosities have been developed for dental applications, mimicking the properties of human bone and promoting bone regeneration [10]. Pore-induced dental implants with enhanced osseointegration have been created using additive manufacturing methods, improving mechanical properties and resolution [11]. Surface modification techniques are also being used to enhance the biocompatibility, osteogenesis, and bactericidal effects of titanium implants [12]. Additionally, bionic micro/nano antibacterial structures have been developed for oral cavity implants, inhibiting bacteria, and reducing periarthritis [13]. FGMs play a pivotal role in dental applications, contributing to the design of root canal posts and implant abutments. By replicating the natural gradient of tooth structure, FGMs enhance the overall strength and longevity of dental restorations. The gradual changes in material properties optimize load distribution, improving the performance and longevity of dental implants. These advancements in FGNMs can potentially improve the performance and longevity of dental implants.

9.2.2 Drug delivery systems

FG structures can be incorporated into drug delivery systems to control the release rate of therapeutic agents. Gradient materials can facilitate sustained and localized drug delivery, enhancing treatment efficacy and minimizing side effects. These structures offer the potential to optimize drug release parameters such as timing, dosage, and site-specific delivery [14]. Micro-electro-mechanical systems (MEMS) have emerged as a solution to challenges in drug release, combining mechanical and electrical components to achieve controlled drug delivery [15]. Tailor-made micro- and nanocarriers have also been developed for drug delivery, offering high loading capacity, trigger release mechanisms, and biocompatibility [16]. Additionally, the use of micro- and nanoencapsulation techniques has advanced drug delivery systems, protecting drugs from degradation and controlling their release [17]. These advancements in FGNMs have the potential to improve therapeutic outcomes and address challenges in drug delivery, such as long-term treatments and individual dosing regimens [18]. The application of FGMs extends to drug delivery devices, particularly in stents and implantable drug pumps. FGM coatings on stents allow for controlled drug release, preventing restenosis in blood vessels. Implantable drug pumps with FGM membranes enable precise control over drug diffusion rates, ensuring a sustained and targeted release over time.

9.2.3 Biosensors

FGNMs can enhance the sensitivity of biosensors by optimizing the surface properties and interfaces. This is crucial for the detection of biomolecules and disease markers in medical diagnostics. These structures, which have varying composition and properties, can enhance the performance of biosensors by improving sensitivity and limit of detection [19]. Graded photonic crystals with defect layers have been investigated for biosensor applications, and the thickness parameter of the graded structure has been found to significantly improve sensor parameters [20]. Hybrid semiconductor structures that integrate nanoelectromechanical systems with metal oxide semiconductor technology have been proposed for biosensing. These structures feature functionalized beams that can detect biomolecules, such as enzymes, bacteria, viruses, and DNA/RNA chains [21]. FGM-based sensors find applications in clinical monitoring, such as wearable devices or implantable sensors. The graded responses of FGMs to physiological changes make them valuable for developing sensors with improved sensitivity and specificity. These sensors contribute to advancements in diagnostic and monitoring technologies.

9.2.4 Tissue engineering

In tissue engineering, FGNMs can be used to design scaffolds with varying mechanical and biochemical properties to mimic the native tissue. This supports cell growth, differentiation, and tissue regeneration. These structures can be used to regenerate or replace damaged tissues and organs by providing scaffolds with specific three-dimensional structures to guide cell activities [22]. Researchers have used techniques such as selective laser melting and fused filament fabrication to fabricate FG scaffolds with tailored properties [23]. The use of triply periodic minimal surface (TPMS) has been proposed as an effective method to achieve functional gradients in scaffold design, allowing for the generation of bone-mimicking architectures [24]. The combination of 3D printing and nanotechnology has led to the development of complex structures with improved mechanical performance [25]. Nanocomposite inks have emerged as a promising approach for producing biofunctional and stimuli-responsive environments in tissue engineering [21].

FGMs are integral in the field of tissue engineering, where scaffolds with graded properties are designed to create a biomimetic environment for cell growth and tissue regeneration as shown in figure 9.3. The customization of FGMs allows for

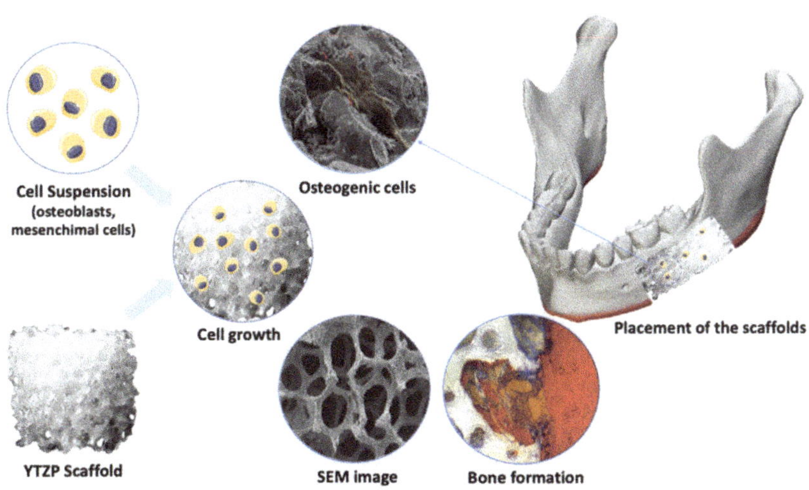

Figure 9.3. Schematics of the cell growth in the porous zirconia structures. Reproduced from [26]. CC BY 4.0.

the development of scaffolds with tailored mechanical and biochemical properties, promoting optimal conditions for cell differentiation and tissue integration.

9.2.5 Prosthetics and wearable devices

Improved comfort and functionality: Gradient structures in prosthetics and wearable devices can provide a more natural feel by matching the mechanical properties of the device with the surrounding tissues. This can improve user comfort and functionality. FGMs are advanced composites with gradually varying composition and microstructure, resulting in enhanced properties. They are used to achieve desired properties in prosthetics [27]. FGMs also find application in complex micro- and nanoelectronic and energy conversion devices [28]. Additionally, FGMs are used in microelectrical discharge machining (microEDM) for the fabrication of 3D microstructures [29]. FGMs can be obtained using spark plasma sintering (SPS) technology, and their mechanical characteristics can be evaluated using Vickers microhardness [30]. Furthermore, FG additive manufacturing (FGAM) allows for the production of freeform structures with customizable properties, making it suitable for prosthetics and wearable devices [31]. FGMs contribute to advancements in prosthetics, both in limb prosthetics and dental prosthetics. In limb prosthetics, FGMs provide adaptive mechanical properties, improving comfort and functionality for users. In dental prosthetics, such as crowns and bridges, FGMs enhance stress distribution, ensuring optimal performance and longevity of the prosthetic devices.

9.2.6 Diagnostic imaging

FGNMs can be employed as contrast agents in imaging techniques such as magnetic resonance imaging (MRI) or ultrasound, enhancing the visibility of specific tissues or structures. These structures have been utilized as contrasting agents and fluorescent materials in various imaging techniques such as MRI, computed tomography (CT), positron emission tomography (PET), near-infrared fluorescence imaging, ultrasound, and photoacoustic imaging [32]. Nanomaterials, produced using nanotechnology, have emerged as excellent materials in imaging applications due to their unique properties like high surface area, surface plasmon resonance, superparamagnetic, and fluorescence [33]. Additionally, the adaptability of optical fiber has facilitated the creation of fiber probes that incorporate optimized waveguide designs and integrated functional materials. This innovation has significantly improved the interaction between optical modes and biological samples, resulting in extremely sensitive biosensors that can detect very low levels of substances. Additionally, by manipulating the dispersion and nonlinearity of light traveling through the fiber core, or by creating metal-coated tapered fiber tips with apertures smaller than the wavelength of light, it has been possible to achieve high-resolution imaging of biological samples [34]. These advancements in FGNMs have the potential to revolutionize biomedical imaging techniques and improve detection limits, sensitivity, optical resolution, imaging depth, and stimulation precision [21, 35]. FGMs are utilized in the development of specialized diagnostic tools, particularly in imaging and diagnostic equipment. The graded properties of FGMs enhance the performance of these tools, improving sensitivity and accuracy in medical diagnostics.

9.2.7 Neural interfaces

FGMs can be used in neural interfaces, such as electrodes and neural probes, to improve biocompatibility and reduce inflammation, enabling better integration with the nervous system. Current advancements in the application of FGNMs in neural interfaces on biomaterials include the use of semiconducting nanomaterials, specifically silicon-based nanostructures, for sensing and modulation applications [36]. Additionally, biomimetic substrates inspired by the architectural organization of the brain extracellular matrix (ECM) have been designed to enhance neural interfaces and dictate neuronal behavior at the cell–material interface [37]. Strategies such as introducing roughness, using 3D scaffolds and hydrogels, and incorporating anisotropic features have been explored to provide topographical cues and recognizable paths for neuron development and functional connections [38]. Furthermore, advancements in two-photon polymerization and remotely reconfigurable dynamic interfaces are paving the way for smart biointerfaces for *in vitro* applications in neural tissue engineering and repair strategies [39]. Other advancements include the development of minimally invasive architectures such as filamentary probes, conformal sheets, open-mesh networks, and distributed material elements for recording and modulating neural activities [40]. The use of nanomaterial-based microelectrode arrays (MEAs) has also shown promise in improving signal detection, neural modulation precision, and biocompatibility in bidirectional *in vitro* brain–computer interfaces (BCIs).

9.2.8 Microfluidic devices

Enhanced fluid control: In microfluidic devices, gradient structures can be utilized to improve fluid control, leading to better accuracy and efficiency in applications such as lab-on-a-chip devices for medical diagnostics. made of bidirectional FG materials, can improve the stability of nanopipes and provide a guide for designing advanced micro/nanofluidic devices for bioengineering applications [41]. The fluidization of particles in a liquid medium using Faraday waves generated by vibrating a metallic plate can help control the distribution of particles in the fluidic medium, allowing for the processing of heterogeneous materials with spatially varying properties [42]. Graded nanoporous structures exhibit excellent strength and good deformability, making them suitable for structural applications [43]. Spiral-type magnetic micromachines can be used as pumps, channel selectors, and active valves for microfluid manipulation, enabling control over the direction and speed of fluid flow [44].

9.2.9 Biocompatible coatings

FGNMs can be employed as coatings on medical devices to improve biocompatibility, reduce inflammation, and prevent adverse reactions with biological tissues. FGNMs in biocompatibility coatings offer several potential benefits. These coatings can improve the biocompatibility of titanium alloys by removing drawbacks such as decreased biocompatibility caused by the formation of an oxygen diffusion layer (ODL) [45]. They can also act as reliable barriers to prevent nickel diffusion from the substrate to the surface, enhancing the cytocompatibility of the coating [46]. Additionally, these coatings can decrease the corrosion rate of zinc-based metals, reduce the release of zinc ions, and promote hydroxyapatite deposition, leading to improved biocompatible and osteogenic properties [47]. Furthermore, incorporating biofunctional elements into nano/microstructure coatings on titanium implants has been shown to enhance bone repair/regeneration and bacterial resistance [48].

9.2.10 Cancer therapy

Gradient structures can be applied in cancer therapy for targeted drug delivery, ensuring that therapeutic agents are delivered precisely to cancerous tissues while minimizing damage to healthy cells. Nanotechnology, specifically nanomaterials, can address the challenges in clinical cancer care, such as lack of tools for early detection, drug resistance, and lack of targeted therapies [49]. Nanoparticles can enhance the efficacy of chemotherapy for invasive cancers. Engineered robots and microrobots are gaining attention for their ability to effectively reach malignancies and penetrate malignant tissues, allowing for targeted therapeutic delivery [50]. Additionally, nanomaterials and nanomaterial-integrated microdevices are crucial for achieving high performance in cancer diagnosis [21]. These materials, with varied porosity, have wide applications as hollow members in biomedical engineering. The application of FGNMs in biomedical devices reflects their versatility in addressing specific challenges and improving overall performance in various healthcare-related areas.

9.3 Conclusions

In conclusion, the exploration of FGNMs in biomedical devices reveals a promising avenue for advancing healthcare technologies. The diverse applications, ranging from orthopedic implants to diagnostic devices, underscore the versatility and adaptability of FGMs in addressing complex challenges in the biomedical field. The controlled variation in composition and structure at the nano and micro scales empowers researchers and practitioners to tailor mechanical, biological, and physicochemical properties with unprecedented precision.

As evidenced by recent breakthroughs, the integration of FGMs holds great potential for enhancing the performance, longevity, and biocompatibility of biomedical devices. This chapter contributes to the growing body of knowledge in the field, providing a comprehensive overview of current research trends and inspiring future investigations. As we advance into an era of personalized medicine and sophisticated healthcare solutions, the application of FGNMs emerges as a key catalyst for transformative developments in biomedical engineering. FGMs exhibit exceptional characteristics that render them highly suitable for a wide range of applications, particularly in the clinical and biomedical fields. Overall, FGMs stand at the forefront of innovative materials, demonstrating immense potential for revolutionizing various aspects of clinical and biomedical applications.

References

[1] Shi H, Zhou P, Li J, Liu C and Wang L 2021 Functional gradient metallic biomaterials: techniques, current scenery, and future prospects in the biomedical field *Front. Bioeng. Biotechnol.* **8** 616845

[2] Najibi A and Mokhtari T 2023 Functionally graded materials for knee and hip arthroplasty; an update on design, optimization, and manufacturing *Compos. Struct.* **322** 117350

[3] Rouf S, Malik A, Raina A, Irfan Ul Haq M, Naveed N, Zolfagharian A and Bodaghi M 2022 Functionally graded additive manufacturing for orthopedic applications *J. Orthop.* **33** 70–80

[4] Dubey A, Jaiswal S and Lahiri D 2022 Promises of functionally graded material in bone regeneration: current trends, properties, and challenges *ACS Biomater. Sci. Eng.* **8** 1001–27

[5] Chen M 2022 Recent advances and perspective of nanotechnology-based implants for orthopedic applications *Front. Bioeng. Biotechnol.* **10** 878257

[6] Ayatollahi M R, Davari M H, Shirazi H A and Asnafi A 2019 To improve total knee prostheses performance using three-phase ceramic-based functionally graded biomaterials *Front. Mater.* **6** 107

[7] Limmahakhun S, Oloyede A, Sitthiseripratip K, Xiao Y and Yan C 2017 Stiffness and strength tailoring of cobalt chromium graded cellular structures for stress-shielding reduction *Mater. Des.* **114** 633–41

[8] España F A, Balla V K, Bose S and Bandyopadhyay A 2010 Design and fabrication of CoCrMo alloy based novel structures for load bearing implants using laser engineered net shaping *Mater. Sci. Eng.* C **30** 50–7

[9] Majzoobi G H, Mohammadi M and Rahmani K 2022 Microstructural examination and mechanical characterization of Ti/HA and Ti/SiO_2 functionally graded materials fabricated at different loading rates *J. Mech. Behav. Biomed. Mater.* **136** 105497

[10] Hou C, Liu Y, Xu W, Lu X, Guo L, Liu Y, Tian S, Liu B, Zhang J and Wen C 2022 Additive manufacturing of functionally graded porous titanium scaffolds for dental applications *Biomater. Adv.* **139** 213018

[11] Dabaja R, Popa B I, Bak S-Y, Mendonca G and Banu M 2022 Design and manufacturing of a functionally graded porous dental implant *Int. Manufacturing Science and Engineering Conf.* (American Society of Mechanical Engineers)

[12] Li J, Zhou P, Attarilar S and Shi H 2021 Innovative surface modification procedures to achieve micro/nano-graded Ti-based biomedical alloys and implants *Coatings* **11** 647

[13] Wang F, Shi L, He W-X, Han D, Yan Y, Niu Z-Y and Shi S-G 2013 Bioinspired micro/nano fabrication on dental implant–bone interface *Appl. Surf. Sci.* **265** 480–8

[14] Ndebele R T, Yao Q, Shi Y N, Zhai Y Y, Xu H L, Lu C T and Zhao Y Z 2022 Progress in the application of nano- and micro-based drug delivery systems in pulmonary drug delivery *BIO Integr.* **3** 71–83

[15] Villarruel Mendoza L A, Scilletta N A, Bellino M G, Desimone M F and Catalano P N 2020 Recent advances in micro-electro-mechanical devices for controlled drug release applications *Front. Bioeng. Biotechnol.* **8** 827

[16] Chesneau C, Sow A O, Hamachi F, Michely L, Hamadi S, Pires R, Pawlak A and Belbekhouche S 2023 Cyclodextrin-calcium carbonate micro- to nano-particles: targeting vaterite form and hydrophobic drug loading/release *Pharmaceutics* **15** 653

[17] Toworfe G K 2019 Incorporating opioids into micro-to nano-structurally optimized silica xerogel controlled release delivery systems prevents abuse *Eur. Sci. J.* **15** 1

[18] Mandracchia D and Tripodo G 2020 Micro and nano-drug delivery systems *Silk-Based Drug Delivery Systems* (London: The Royal Society of Chemistry) pp 1–24
[19] Shabana Y M, Samy M A, Abdel-Aziz M A, Hindawi M E, Mosry M G, Albarawy A-R M, Omar M M, Mohamed A A and Attia A A 2021 Enhancing the performance of micro-biosensors by functionally graded geometrical and material parameters *Arch. Appl. Mech.* **91** 2497–511
[20] Ankita , Bissa S, Suthar B and Bhargava A 2022 Graded photonic crystal as improved sensor for nanobiophotonic application *Macromol. Symp.* **401** 2100319
[21] Fattahi A M, Sahmani S and Ahmed N A 2020 Nonlocal strain gradient beam model for nonlinear secondary resonance analysis of functionally graded porous micro/nano-beams under periodic hard excitations *Mech. Based Des. Struct. Mach.* **48** 403–32
[22] Guo W, Jiang Z, Zhang C, Zhao L, Jiang Z, Li X and Chen G 2021 Fabrication process of smooth functionally graded materials through a real-time inline control of the component ratio *J. Eur. Ceram. Soc.* **41** 256–65
[23] Hugenberg N R, Dong L, Cooper J A, Corr D T and Oberai A A 2020 Characterization of spatially graded biomechanical scaffolds *J. Biomech. Eng.* **142** 1–11
[24] Zhang X Y, Yan X C, Fang G and Liu M 2020 Biomechanical influence of structural variation strategies on functionally graded scaffolds constructed with triply periodic minimal surface *Addit. Manuf.* **32** 101015
[25] Moheimani R and Dalir H 2020 Static and dynamic solutions of functionally graded micro/nanobeams under external loads using non-local theory *Vibration* **3** 51–69
[26] Resende-Gonçalves C I, Sampaio N, Moreira J, Carvalho O, Caramês J, Manzanares-Céspedes M C, Silva F, Henriques B and Souza J 2022 Porous zirconia blocks for bone repair: an integrative review on biological and mechanical outcomes *Ceramics* **5** 161–72
[27] Saad A 2022 A review on functionally graded materials and their applications in the field of prosthetics *ERJ. Eng. Res. J.* **45** 553–60
[28] Gao Y, Xie Q, Gao T, Yang W, Chen Q, Tian Z, Li L, Liang Y and Wang B 2023 Design of functionally graded Ti–Al alloy with adjustable mechanical properties: a molecular dynamics insights *J. Mater. Res. Technol.* **23** 258–67
[29] Liu W, Li Y and Xu B 2023 Introduction to electrical discharge machining micro/nano structures *Fabrication of Micro/Nano Structures via Precision Machining: Modelling, Processing and Evaluation* (Singapore: Springer Nature) pp 165–75
[30] Shichalin O O *et al* 2023 Functionally gradient material fabrication based on Cr, Ti, Fe, Ni, Co, Cu metal layers via spark plasma sintering *Coatings* **13** 138
[31] Pei E, Kabir I R, Godec D, Gonzalez-Gutierrez J and Nordin A 2021 Functionally graded additive manufacturing *Additive Manufacturing with Functionalized Nanomaterials* (Amsterdam: Elsevier) pp 35–54
[32] Suchetha A D 2023 Current advancements in biomedical research *RGUHS J. Dent. Sci.* **15** 6
[33] Yamamoto T, Yoshiya M and Nhat H N 2023 Recent progress in nanostructured functional materials and their applications II *Mater. Trans.* **64** MT-M2022181
[34] Thirugnanasambandan T 2021 Nanoparticles for biomedical imaging advancements *Nanoparticles in Analytical and Medical Devices* (Amsterdam: Elsevier) pp 127–54
[35] Yu X, Zhang S, Olivo M and Li N 2020 Micro- and nano-fiber probes for optical sensing, imaging, and stimulation in biomedical applications *Photonics Res.* **8** 1703
[36] A Miao B and Tian B 2023 Nanomaterial-enabled bioelectrical interfaces *Encyclopedia of Nanomaterials* (Amsterdam: Elsevier) pp 469–83

[37] Mariano A, Bovio C L, Criscuolo V and Santoro F 2022 Bioinspired micro- and nano-structured neural interfaces *Nanotechnology* **33** 492501
[38] Chen J and Wu S 2022 Advanced architectures and materials of functional devices for neural interfaces *Highlights Sci. Eng. Technol.* **23** 168–76
[39] Kim S *et al* 2023 Materials and structural designs for neural interfaces *ACS Appl. Electron. Mater.* **5** 1926–46
[40] Liu Y, Xu S, Yang Y, Zhang K, He E, Liang W, Luo J, Wu Y and Cai X 2023 Nanomaterial-based microelectrode arrays for *in vitro* bidirectional brain–computer interfaces: a review *Microsyst. Nanoeng.* **9** 13
[41] Lyu Z, Tang H and Xia H 2023 Thermo-mechanical vibration and stability behaviors of bi-directional FG nano-pipe conveying fluid *Thin-Walled Struct* **188** 110784
[42] Kumar D, Kumar D and Tigga A M 2022 A novel method for fabricating functionally graded materials by vibration-assisted casting *Fabrication and Machining of Advanced Materials and Composites* (Boca Raton, FL: CRC Press) pp 205–21
[43] He L and Abdolrahim N 2022 Mechanical enhancement of graded nanoporous structure *J. Eng. Mater. Technol.* **144** 011007
[44] Ji D M and Kim S H 2019 Functional fluid-manipulation using spiral-type magnetic micromachines as micropumps, active valves, and channel selectors *IEEE Access* **7** 145596–603
[45] Cheraghali B, Ghasemi H M, Abedini M and Yazdi R 2022 Functionally graded oxygen-containing coating on CP-titanium for bio-applications: characterization, biocompatibility and tribocorrosion behavior *J. Mater. Res. Technol.* **21** 104–20
[46] Baigonakova G A, Marchenko E S, Yasenchuk Y F, Kokorev O V, Vorozhtsov A B and Kulbakin D E 2022 Microstructural characterization, wettability and cytocompatibility of gradient coatings synthesized by gas nitriding of three-layer Ti/Ni/Ti nanolaminates magnetron sputtered on the TiNi substrate *Surf. Coat. Technol.* **436** 128291
[47] Qian J *et al* 2022 Micro/nano-structured metal–organic/inorganic hybrid coatings on biodegradable Zn for osteogenic and biocompatible improvement *Adv. Mater. Interfaces* **9** 2101852
[48] Xu N, Fu J, Zhao L, Chu P K and Huo K 2020 Biofunctional elements incorporated nano/microstructured coatings on titanium implants with enhanced osteogenic and antibacterial performance *Adv. Healthcare Mater.* **9** 2000681
[49] Swarna latha B 2020 Nano world in cancer therapy *Asian Pac. J. Cancer Biol.* **5** 183–8
[50] Zheng S-Y 2021 Develop Micro/Nano Technologies for Cancer Diagnosis *2021 IEEE 34th Int. Conf. on Micro Electro Mechanical Systems (MEMS)* (Piscataway, NJ: IEEE) pp 346–9

www.ingramcontent.com/pod-product-compliance
Ingram Content Group UK Ltd.
Pitfield, Milton Keynes, MK11 3LW, UK
UKHW051851100926
8556IPUK00005B/571

9 780750 360227